Easter Island

Scientific Exploration into the World's Environmental Problems in Microcosm

Easter Island

Scientific Exploration into the World's Environmental Problems in Microcosm

Edited by

John Loret
Science Museum of Long Island
Plandome, New York

and

John T. Tanacredi
Dowling College
Oakdale, New York and
American Museum of Natural History
New York, New York

Kluwer Academic / Plenum Publishers
New York, Boston, Dordrecht, London, Moscow

Library of Congress Cataloging-in-Publication Data

Easter Island : scientific exploration into the world's environmental problems in microcosm / edited by John Loret and John T. Tanacredi.
p. cm.
Includes bibliographical references and index.
ISBN 0-306-47494-8
1. Environmental management--Easter Island. 2. Environmental protection--Easter Island. 3. Oceanography--Easter Island. 4. Marine biology--Easter Island. I. Loret, John. II. Tanacredi, John T.

GE320.E19E27 2003
333.7'2'099618--dc21

2003051587

ISBN 0-306-47494-8

233 Spring Street, New York, New York 10013

http://www.wkap.nl/

10 9 8 7 6 5 4 3 2 1

A C.I.P. record for this book is available from the Library of Congress

Printed in the United States of America

Dedication

Dedicated to the memory of Dr. Thor Heyerdahl
Respected, esteemed and admired for years by many.
Thor Heyerdahl, a scholar, an intrepid explorer,
a natural leader and humanitarian.

Co-Editor
John Loret, Ph.D. D.Sc.

Dedicated to young explorers everywhere who may keep the passion always alive in their quest for knowledge; their compassion for all humanity and their appreciation of all life on earth.

Co-Editor
John T. Tanacredi, Ph.D.

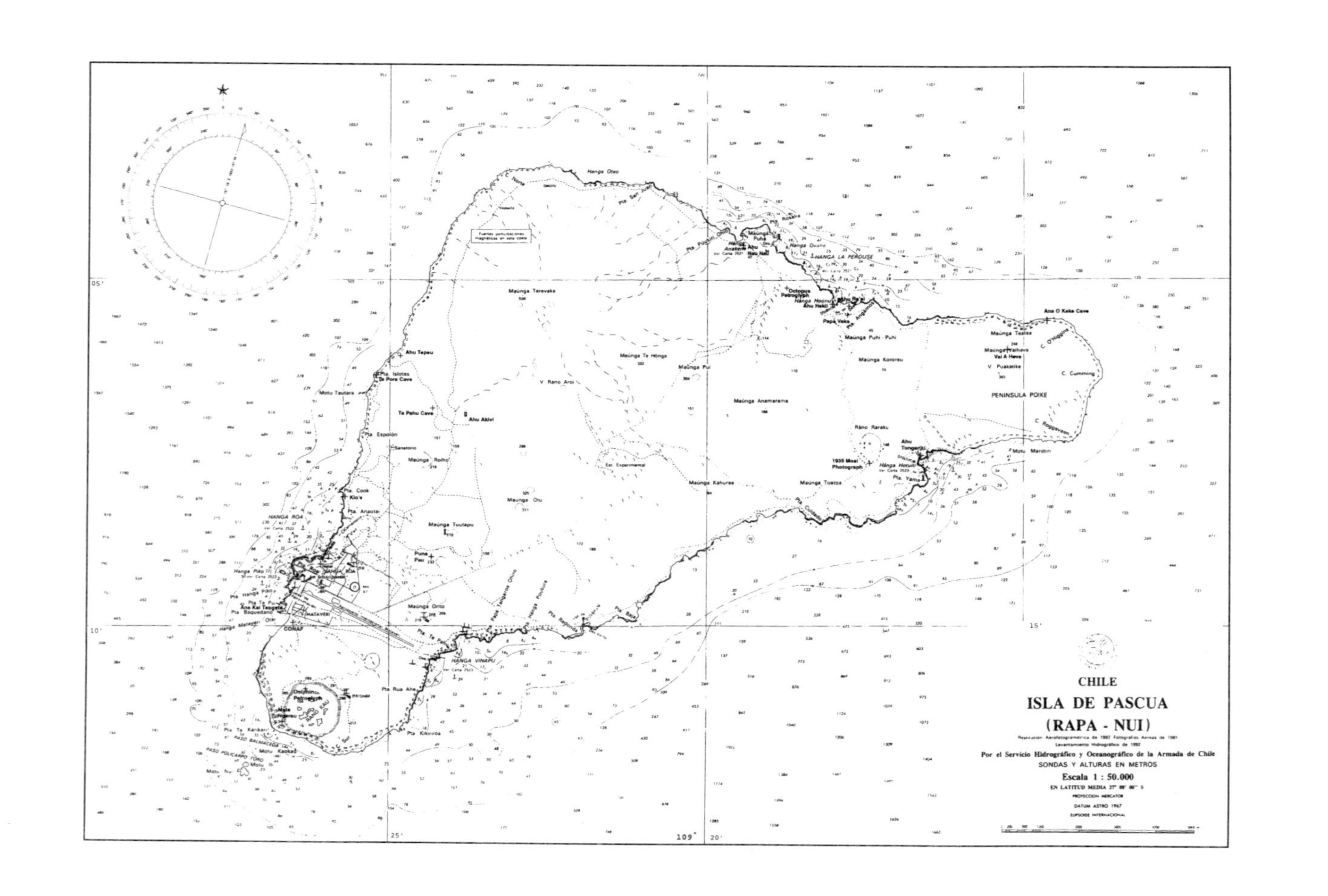
CHILE
ISLA DE PASCUA
(RAPA - NUI)
Levantamiento Hidrográfico de 1992
Por el Servicio Hidrográfico y Oceanográfico de la Armada de Chile
SONDAS Y ALTURAS EN METROS
Escala 1 : 50.000
PROYECCION MERCATOR
DATUM ASTRO 1967
ELIPSOIDE INTERNACIONAL
PENINSULA POIKE
HANGA LA PEROUSE
HANGA ROA
HANGA VINAPU
Hanga Oteo
Maúnga Terevaka
V. Rano Aroi
Maúnga Te Hónga
Maúnga Anamarama
Ráno Raraku
Maúnga Puhi - Puhi
Maúnga Kororeu
Ana O Keke Cave
C. O'Higgins
C. Cumming
C. Roggeveen
Motu Marotiri
Ahu Tongariki
1935 Moai Photograph
Hanga Hotuiti
Pta. Yama
Maúnga Toatoa
Pta. Cuidado
Maúnga Kahurea
Ahu Tepeu
Pta. Islotes
Te Pora Cave
Motu Tautara
Te Pahu Cave
Ahu Akivi
Pta. Espolón
Maúnga Roiho
Pta. Cook
Pta. Anaotai
Maúnga Tuutapu
Maúnga Otu
Est. Experimental
Puna Pau
Maúnga Orito
MATAVERI
CONAF
Pta. Baquedano
Pta. Rua Ahe
Pta. Kikiroa
PASO BALMACEDA
PASO POLICARPO TORO
Motu Kaokao
Motu Nui
Mata Ngarau
Octopus Petroglyph
Papa Vaka
Ahu Heki'i
Hanga Ovahe
Anakena
15'
20'
25'
109°
05'
10'

Foreword

MARC A. KOENINGS
General Superintendent, Gateway National Recreation Area and New York Harbor Parks

I have been in the National Park Service for over 25 years now, and one of the most enduring legacies of my tenure as a Superintendent is the work done by thousands of dedicated people committed to preserving our natural and cultural heritage. This dedication is no less intense on the international scaled even with what sometimes appears to be overwhelming obstacles.

My first reaction upon picking up the draft manuscript was — I hope this work is not another eco-tour attempt playing on the lack of information available, on the mysteries of Easter Island. I also wanted to know...what was so compelling to engage the National Park Service? From several different facets, it is clear that this World Heritage Site is still, after over 50 years since Jacque Cousteau's work on the island, a fascinating area to explore and learn about a culture that has only remnants remaining, while documenting a marine ecology still mostly unknown.

Primarily, what are the very real lessons we can learn from Easter Island as we daily ponder the impacts of global climate change? It is becoming increasingly evident that we live in a closed system; it may be on a global scale, but it is a closed system. We, like the past residents of Easter Island, are pushing the limits of resource utilization that will result in the failure of the health of ecosystems and ultimately a decline in human populations. Easter Islanders had an option to migrate to another region; they didn't — a collapse of the global system does not afford that luxury for us; we can't...

The issue of tourism offers simultaneously a curse and a blessing. It is presented with some very practical management suggestions to protect the integrity of the Moai — perhaps not unlike what is done at Stone Henge. Equally practical are suggestions on how to map the Moai placement and most importantly how to protect the stone from further deterioration from the ravages of tourists and nature.

Equally compelling is the discussion on the contemporary society. What are their dreams and aspirations? Can the destiny of the Rapa Nui include that elusive balance of modern convenience with the preservation of their remarkable heritage?

It has been a wonderful experience for me over the years to work and help shape the direction of our nation's National Parks. I am very happy to be able to make a contribution to our global thinking on environment and culture by presenting this work to you...truly a contribution to our world heritage. I thank Dr. Loret and Dr. Tanacredi for making it happen.

Photo images 64 years apart exhibit changes in vegetative cover and surface erosion to Moai figures.

Acknowledgments

This work is the result of the combined efforts of all those 1996 through 2000 Easter Island Expedition participants listed at the end of this work. An enthusiastic "bravo" and "thank you" go out to all their efforts and contributions. Special appreciation goes to the staff of the Explorers' Club, The Science Museum of Long Island, the National Park Service at Gateway National Recreation Area for all their assistance, funding support and patience in making this book possible. We would care to acknowledge all those who took the time from their own hectic schedule to review the draft manuscript for this effort especially Dr. P.A. Buckley, Dr. Stephen Shafer, Mr. Fred Rubel, Dr. Chikashi Sato, Dr. Robert Cook, Dr. Paul V. Loiselle, Dr. William Kornblum, Dr. George Fame, Mr. Marc Koenings. Dr. Lori Zaikowski and Dr. Martin P. Schreibman; without your professional eye to detail and accuracy this work could not have been completed. Special thanks go to all the peoples of Easter Island and Chile; especially Sergio Rapu, Jose Migual and CONAF (Chilian National Park Service). We would care to acknowledge those who grappled with the preparation of the manuscript, initially Debbie Wynne and finally Debra Jedlicka, both of Dowling College; they performed beyond the call of duty. Finally to Ms. Joanna Lawrence, Editor at Kluwer Academic/Plenum Publishers, for her foresight, intuition and guidance in making this work happen. It should be noted that any mistakes or omissions are totally the results of the editors.

One of the thousands of petroglyphs on the island depicting an octypus … species of a size long gone from the island's coastline.

Contents

Introduction: An Easter Island Experience—How it all Started

JOHN LORET
Director, The Science Museum of Long Island; Past President, The Explorers Club

1. BACKGROUND

In the fall of 1954 as a graduate student at the University of Oslo, Norway. I wanted to escape the cold, dark rain season in Norway so I asked to be sent to the Canary Islands to collect algae specimens while skin-diving into under-sea caves along the coast. I flew from Oslo to the City of Las Palmas on the island of Grand Canary. Outside the city I set up camp on the beach at a beautiful site. On the second day while I was in the sea diving, children stole my tent, money, stove and equipment. Fortunately, the clothes, sneakers and my skin diving equipment, items I was wearing, remained. Before I left Oslo, my professor, Dr. Braarud told me that Thor Heyerdahl was staying at the Hotel Santa Catalina in Las Palmas with his wife Yvonne and new baby girl Annette.

Since his remarkable expedition on the Kon Tiki raft a few years before, Heyerdahl and his crew were my heroes. Not having any friends or contacts in Las Palmas, I decided to seek help from Heyerdahl. I remember feeling a bit awkward and apprehensive as I approached the lobby of the hotel. I called his room and was told to wait in the lobby. He was a tall, gentle and soft-spoken man, with the bluest eyes I have ever seen. He listened quietly to all I had to say. He was interested in my work with the university and asked if he could go out with me to dive. After lunch he provided me sufficient funds to rent a small room in town. I also wired my roommate in Norway, Thomas Larner who sent additional money.

The next day we rented a boat with a fisherman and went out beyond the surf.

The boatman was careful not to come too close to the breaking waves. The lava caves are along the wall of land extending into the sea. We donned our snorkel gear, I had my net bag with collecting vials as well as my Lica camera mounted in a waterproof metal housing. I dove down about 35' into a cave and with the vial scraped the algae off the cave ceiling. On the surface I labeled the vial with a wax pencil and placed it in my net bag. We were talking at the surface when a seven foot white tip shark came up close to us. The fisherman became worried and started loudly and rapidly speaking in Spanish. Thor and I dove under water, swam close up to the shark who then turned and swam away. However, I did manage to take his picture while he was still in view. Later, Thor told me that this was the first encounter he had with a shark underwater.

2. GOING ON THE EXPEDITION

In the summer of 1955 I was working for the United States occupation forces in Austria. My assignment was to set up a camp in the mountains for the children of the American troops stationed in there. A week before camp was to open, Austria received her independence and all U.S. forces were evacuated. I was out of a job. One evening I was reading the newspaper, there was an article describing Dr. Heyerdahls' new archaeological expedition planned for Easter Island and the Society Islands. It was starting in September and to last ten months or more. This was my chance! I was free to travel; my work at the university could wait. So I called Thor Heyerdahl in Oslo to ask if there was a position for me on the expedition. He said there is an opening for a seaman and diver. He said that with my biology background I could also assist the expedition doctor and surgeon in emergencies. So back to Oslo I went.

The ship selected for the expedition was a 150-foot trawler; a most seaworthy ship equipped for anything the ocean could muster. It was the "Christian Bjeland", built to work in the seas off Greenland (Fig. 1).

Figure 1. Christian Bjeland

On a rainy September day, H.R.H. Crown Prince Olav (later King Olav) had agreed to be patron of the expedition. He came aboard to meet the crew and wish us all well. The "Christian Bjeland" sailed with her crew across the Atlantic and into the Caribbean to Christbal, Panama. In route we sailed directly into hurricane "Carol". For days we sat on our nose, no meals could be cooked and many were seasick. However the "Christian Bjeland" rode the storm without difficulty. At Christbal we took aboard new supplies along with Dr. Heyerdahl, his wife Yvonne, his daughter Annette and the scientists, Dr. William Mulloy, Dr. Carlyle Smith, Dr. Edwin Ferdon from the United States, Anne Skjolsvald from Norway and Gonzalo Figueroa from Chile. After sailing through the Panama Canal we set our course for the Galapagos Islands.

Here I had the opportunity to try our new diving gear, the Scott Hydro Pack with two 72 cubic foot tanks. I used the Norwegian Viking dry suit. It was interesting diving with sea lions and marine iguanas. The sea was full of life everywhere – groupers, snappers, jacks and even the giant manta ray. It was like diving in an aquarium. We left our mail in a wooden barrel in Targus cove for the next ship bound for a homeport. Whalers who would be gone on expeditions for years set up this system. Ships would also stay at the Galapagos to take on stores of wild goats for fresh meat. These were brought to the island where they could multiply and be a constant source of fresh meat. Unfortunately, Mariners would also capture the giant tortoises, which could be kept alive for months in the ship's bilge.

We then set our course for the uninhabited islands of Sala y Gomez, which are located 600 miles east of Easter Island. The islands rocky coastline and heavy seas made it difficult to land a boat. The Captain, Arne Hartmark and First Mate Sanny, lashed a raft together with two aluminium pontoons and timbers. The raft was put overboard. With a sextant and chronometer, the Captain and I went aboard and began rowing towards shore. The raft secured with a long line to the ship. As we came close to shore I jumped overboard with snorkel gear to guide the raft between rocks and coral heads. The water was alive with fish. Sharks, rays, butterfly fish and tangs were just a few seen as I swam alongside the raft. The Captain took his bearings and we returned to the ship.

3. FATHER SEBASTIAN ENGLERT

Father Sebastian was a Catholic capuchin priest from Austria. In 1955 he had been on Easter Island for twenty-five years where he planned to remain until his death. He was the most powerful man on the island. He had written a book on Easter Island, as well as a dictionary of the native Rapa Nui language. He could speak to the natives in their own language and was fluent in English, Spanish, French and German. He loved the native people and looked upon them as his children.

I was assigned to work with Dr. Carlyle Smith at Te Pito Kura in La Perouse Bay. One day Father Sebastian came up to me and said, "Loret, that's a Christian name, you must be Catholic. What are you doing here with these Norwegian heathens?" I thought for a long moment. It was true. I was raised a Roman Catholic and as a youngster I was an alter boy in St. Michael's Parish in Brooklyn, New York where we also had capuchin priests. I replied that "I used to be Catholic, Father, but haven't practiced the religion for many years." "Nonsense," he said, "once you are a Catholic you're always a Catholic." We talked for a while and he then asked if I would be interested in serving mass for him. I told him that I had forgotten most of my Latin. "We will take care of that," he said. "I'll talk to Heyerdahl" and then he left.

Later that evening, Thor approached me saying that this would be a good thing for the expedition if I worked with Father Sebastian. I would be given time off to ride into Hangaroa and stay with Father Sebastian on Saturday and Sunday. Father Sebastian likes to drink Scottish whiskey. Thor gave me a bottle of Johnny Walker scotch for Father Sebastian. After a wonderful dinner of chicken, Father Sebastian brought out two glasses and poured out the scotch. We talked about many things, conditions of Europe, life on Easter Island, his love for the natives until the bottle was empty. Father

Sebastian had to fast before midnight so we then went to bed. About 6:00 am the roosters were crowing and Father Sebastian woke me up. "We must say mass first at the leper colony." I had no idea there were lepers on Easter Island. I quickly dressed and followed Father Sebastian and three nuns outside of town. There must have been twenty lepers with various stages of the disease. Father Sebastian set up his altar on a table outdoors and started mass. I fumbled through the Latin the best I could with Father Sebastian completing segments that I had forgotten. All of the lepers received communion except me. The mass ended and we returned to town to Church for the main mass.

Figure 2. Leper Colony

The church was small but beautiful. With the altar at the far end and a thatched roof, the sides and front were completely open except for a small stonewall. Father Sebastian would say mass in Rapa nui, the natives would sing Rapa nui songs, and birds would fly in and out of the church also singing. It was one of the most beautiful scenes I can ever remember. Of all the negative things I've heard and read about missionaries, Father Sebastian was indeed an exception. He loved his people and they loved and trusted him. There wasn't a person on the island who wouldn't offer to assist him. The natives even selected him to negotiate terms for diggers who were to work with Heyerdahl.

Figure 3. Father Sebastian and me in front of the old church

Today the church is much larger, has walls, a metal roof, no birds can fly in and out and the Mass is said in Spanish. The older natives still sing their Rapa Nui songs. Father Sebastian is buried in a grave on the south side of the church.

4. EASTER ISLAND

It was late afternoon in October 1955 as we quietly sailed into HoTuiti, a small cove located on the southeastern corner of Easter Island. The massive Poike Peninsula was off to starboard. We anchored but the Captain was not happy with the bottom. There are very few good anchorages off Easter Island. The sea bottom is rocky and where there is sand and a clear bottom there are coral heads. If you anchor in clear sand the ship moves and the anchor chain snags around the coral heads. Many ships visiting Easter Island have lost their anchors forcing them to leave sooner than expected. The constantly changing wind makes it difficult for a vessel to find a safe anchorage under a lee shore. As the diver for the expedition I envisioned my work would be for the future; untangling anchor chain to free anchors. This we did on several occasions.

From the deck of the ship we could see the giant statues standing along the slopes of volcanic crater of Rano Rakau. That evening we had native male visitors that came aboard to trade. They were a happy people, smiling and friendly but poorly dressed. Hidden under their clothing were woodcarvings. After some trading one man played the guitar. He played native music while another man sang along. Then we all began to dance.

In the morning we weighed anchor and sailed around the Island. The north side of the Island was the calmest while the southern shore had the heaviest surf. The southern coast is the most exposed to the southern ocean. There is nothing between Easter Island and the Antarctic Ocean. Years later we would find that this affected the species of marine organisms found along the southern shore and those of the north coast.

Figure 4. Easter Island southern coast

There were no trees to be seen except as we approached the village of Hangaroa, which had a few scattered Eucalyptus trees. Eucalyptus trees were imported probably from Australia because of their fast growth. Unfortunately, the Eucalyptus drains the soil of all moisture leaving the soil barren for other species to establish. Today there are large stands of these trees in many locations on the island. In the 1960s a large grove of coconut palms were planted at Anakena beach. These trees have taken root and are reseeding.

We selected Easter Island's only beach to establish a campsite. It is a beautiful sandy area on the north side of the island. Small two-man tents for the crew and scientists were set up. The tents had cots and were quite comfortable. A large dining tent was used for eating and recreation. Recreation consisted of reading, listening to music and of course, lively conversations.

Figure 5. Our campsite at Anakena

The many months we spent on Easter Island provided endless adventures and challenges i.e.: exploring ahus, moais and rock carvings over the Island, crawling into caves, diving and spearing fish in the sea, scuba diving to recover anchors and discovering sunken shipwrecks. Working with scientists such as William Mulloy, Carlyle Smith, Arne Skolsvold and Edward Ferdon was better than attending graduate school. In fact, it caused me to consider changing my field of marine biology to archaeology. Later, I did return to complete my work at the University of Oslo and later finished my Ph.D. in Environmental Science.

With all the mystery surrounding Easter Island I found the most interesting to be the natives, making friends and exploring with them. We traveled over the entire Island on horseback, which was quite an adventure. The locals had little or nothing of any value except their woodcarvings used for payment. They would be grateful if we shared a lunch with them, or caught them a lobster, or gave them an old shirt or pants. Many times I thought to myself what will happen to these happy people? In 1955 there was only one ship a year that visited the Island. With the exception of a few Europeans and Chileans including the small naval garrison, the entire population was Easter Islanders. All the people spoke Rapa Nui the native language and in school Spanish was the second language. The Island population at that time was only a bit over 900. The natives had a deep respect for their Island and demonstrated great pride when we discovered a petroglyph, cave or any artefact made by their ancestors.

Figure 6. Some of the ship's crew

The entire Island had been a sheep farm for over 100 years ending about 1963. At the time, the population had grown to over 2500 people comprising 50% immigrants mostly from Chile and others from Germany and some from Scandinavia. In 1968 NASA was seeking an alternate landing field for the space shuttle in the South Pacific and constructed a huge landing field capable of receiving the largest of modern jets on Easter Island. Now tourist and immigrants could easily get to the Island. In forty years this island community made a shift from a primitive way of life into a modern society. The island now has hotels, restaurants, a bank, hospital, schools, bars, supermarkets, discotheques, auto service station, car rental agencies and other services related shops thus, the society has changed to a cash economy. Tourism is now the Island's main source of income. Over 10,000 tourists arrive each year. There are four scheduled flights a week. Immigrants, Islanders and tourists can fly to the South American mainland, or go on to the islands to the east such as Tahiti and Samoa. Small cargo vessels now visit the island four times a year and Cruise ships come in several times a year depending on the weather. All this growth has had adverse effect on the Easter Island people. Today Spanish and English are taught in the schools, consequently less than 10% of the children can speak Rapa Nui. The friendly natives of 1955 are now too busy to take time to show visitors the Island except for money. The Island I remember in 1955 is gone forever. A new aggressive society has taken its place.

Figure 7. John Loret at Rano Rakau

The impact of increased tourism is one of the most critical problems affecting Easter Island today. The present boundaries of Easter Island National Park a World Heritage site, covers about 40% of the Island's surface area. These limits were established in 1976 and were fixed to encompass most of the monuments including: ahus, moai, caves, quarries and ceremonial centers, all rock paintings and most of the estimated 5000 petroglyphs. Because of the vast number of sites and distances to travel to see them the situation has reached a critical level. Visitors indiscriminately climb over statues and ahus and walk on petroglyphs.

The tourists are not the only problem to contend with. The native owned cattle and horses are not fenced in and are constantly grazing and wandering over many of these monuments. The animals weight breaks or upturns ahus stones. Grasslands are burned to provide fresh grass shoots for the cattle. The heats from these fires have caused stones to crack or shatter. Easter Island has a management plan, however, it is insufficient to cope with the vast problems. In 1998 and 1999 our expedition included a team from the United States National Park Service to work with the Chilean counterpart "CONAF" to establish some guidelines for a National Park management plan. The U.S. National Park system has a great deal of experience in the conservation of national monuments and working with massive numbers and visitors to parks.

5. STARTING AN EXPEDITION

While Director for Environmental Studies at Queens College City University of New York, I developed an interdisciplinary approach to study environments that had been the topic of my doctoral dissertation. In 1991, I re-visited Easter Island while lecturing aboard a Cruise ship. It was completely amazing to see how the Island had changed since 1955. Cars and taxis were lined up to take tourists on tours of sites and old horse trails were replaced with roads. Self-appointed guides were competing with each other to show their Island's treasures. I rented a jeep and driver and went off to see as much of the Island as I could, visiting vaguely remembered sites. I had perused much of the work done on Easter Island after 1955-56 by researchers such as: C. Smith, W. Mulloy, P.C. McCoy, W.S. Ayres, J. Randall, J. Flenley and others. I realized that there is still much to be studied and that I wanted to be a part of it.

As Executive Director for the Science Museum of Long Island and President of the Explorers Club I decided to put together an inter-disciplinary scientific team to work on Easter Island. From 1969-1978 during expeditions to the Yucatan, Chiapes and Quintano Roo, we had a team comprised of a carbonate sedimentologist, marine biologist, ecologist and archaeologist. The combined efforts and input of each specialist allowed for greater understanding of how the Mayan culture functioned. We used this model in other environments including Iceland, the Amazon Rain Forest and the Virgin Islands. A unique and highly qualified team was formed and we started work in August 1996. We returned to Easter Island each year until 1999. Although work on Easter Island will always be "ongoing", we are presenting in this volume the work completed to and into 2001.

6. 1997 EXPEDITION

Our research on Easter Island presents the most challenging opportunity we have undertaken. The preliminary work done by Loret and Hemm in 1996 stimulated our interest to organize a more comprehensive expedition mobilizing several scientific disciplines for August 1997, 1998 and 1999.

Much like that of the Mayans, Easter Island's culture, living within a far more limited and fragile land mass, had fallen for many of the same reasons: overpopulation (with no other place to go), environmental degradation, warfare among the clans for diminishing resources and arable land, and, instead of drought, perhaps as a result of a prolonged El Nino.

In August 1997, along with expedition photographers and explorer, Robert Hemm and Marcello Mendez, I returned to Easter Island with an interdisciplinary scientific team. Our objective was to reconstruct events in time that led to the fall of this once great culture. At that time we did not realize that our work would take two to three additional expeditions.

The first settlers who stepped ashore some 1,600 years ago found themselves in a pristine paradise. What then happened? The objective of our expedition hopefully would unveil new evidence in understanding the chronological sequence of ecological changes that occurred over this period of time. From cores obtained by drilling in swamps of extinct volcanic craters and analysis of pollen seeds and twigs with radiocarbon dating, we can determine the vegetation and forest composition back in the time well before human occupation. We can then develop a time-line progression of plants and trees that became extinct due to over harvesting. Easter Island is among the most extreme examples of forest destruction anywhere in the world.

Our focus in 1997 included the following:

Paleontology – Led by Dr. Daniel Mann, Professor Dept. of Geology, University of Alaska and Dr. Dorothy Peteet, Larmont-Doherty Laboratory, Columbia University. A previous study by Dr. John Flenley was done to recreate the vegetation of Easter Island back 30,000 years using pollen analysis taken from cores extracted from swamps in extinct volcanic craters on Easter Island. With our study, we planned a more comprehensive analysis by first taking more and deeper cores to the bottom of the swamps. We then sampled selected intervals very close together (high resolution) for pollen, macrofossils (seeds, needles, twigs), and charcoal identification. Macrofossils are extremely important for three reasons. 1) They can usually be identified to species level and therefore narrow down the ecological niche represented. 2) They also unequivocally establish a species presence at a site. 3) Selected macrofossils can then be used for accelerator mass spectrometry (AMS) radiocarbon dating to give us a precise timing of changes in vegetation and climate through time. We managed to drill deeper than any previous coring, perhaps into sediments older than 60-80,000 years before the present.

Archaeology - Under the direction of archaeologist Sergio Rapu, founder and former Director of the Museum of Archaeology and Ethnology, Hangaroa, Easter Island, we began the reconstruction of Ahu Hanga Piko close to the village of Hanga Roa.

Climatology – Led by Professor David Mucciarrone with Dr. Dunbar of the Dept. of Geology and Environmental Sciences, Stanford University. Using a surface three-inch coring device mounted with a compressor in a boat with divers we were able to obtain cores of heads up to six feet in length. Corals have growth rings similar to those of trees that can be used to date the past events (dendrochronology) and thus provide us with important clues about the past climatologically conditions as yearly water temperature, salinity and approximate rainfall. These cores can also tell us about climatological changes brought by El Nino in past times upon such island groups thus characterizing with more certainty the paleoclimate of Easter Island. Finally, by comparing this data against the archaeological record and paleoecological study, it may explain some of the earlier and later cultural events that occurred on Easter Island, and during which years it was most favorable to travel eastward from Central Polynesia or westward from South America.

Geophysics – Drs. Warren Back and Robert Burr of the Dept. of Physics, University of Arizona have analyzed coral pavement nodule samples taken from the pavement plaza of the Ahu at Anakana for radiocarbon dating. They have found that two samples dated to BP 726 $\pm$ 48 years and 778 $\pm$ 58 years. These two ages are well within errors of each other and appear to be consistent with our hypothesis that they were collected from the sea for the purpose of making pavement. These ages however, are not the true calendar dates. As corals grow they accumulate carbon from seawater. The circulation of the oceans is constantly mixing old carbon with new carbon, thus giving us a correction factor of 400 years. To complicate matters further, in order to obtain a calendar date from the radiocarbon age, we then must use what is called tree ring calibration. This calibration has been derived from counting annual rings from very long lived trees, often the sea water correction and tree ring calibration the two radiocarbon ages of the coral nodules were found to be 1444-1665 B.P. to 1434-1648 B.P. Dr. Beck is also working on dating the coral eyes from the statues at Anakana.

7. 1998-99 EXPEDITIONS

This year we organized a team around our co-sponsors, the National Park Service at Gateway National Recreation Area, coordinated by Dr. John T. Tanacredi, co-leader of the 1998 expedition, who were to survey the near shore marine environment using divers and groups collecting invertebrates in tidal pools. This is a brief summary of their work accomplished. Dr. John T. Tanacredi, co-leader and chief of National Resources for the U.S. National Park Service, collected marine invertebrates in tidal pools along the entire rugged marine shoreline. Collections have been identified and keyed out by taxonomists of the American Museum of National History and Woods

Hole Oceanographic Institution. Several new species of invertebrates were identified.

Dr. Dennis Hubbard, marine geologist funded by the NPS, with five divers conducted an ecological underwater reconnaissance of the Easter Island sub-tidal marine environment to depths of 130 feet. This is the first comprehensive underwater survey ever completed for Easter Island. Easter Island is in the sub-tropic region 27 degrees 10' south latitude 109 degrees 20' west longitude, and well south of where you would expect to find coral reefs. The diving team did discover a small healthy reef off the southwest promontory of Rano Kau. This is the first reported live coral reef for Easter Island.

Dr. Blaine Cliver, U.S. National Park Service, using photogrammatry is determining the rates of erosion of petrographs, and eventually damage due to recarving and human and animal (cattle and horse) activity. Dr. Cliver also measure moai and ahu sites in order to estimate the rates of weathering and human impact.

Dr. Daniel Mann, soil geologist, University of Alaska, obtained cores from the crater lake in Rano Rakaku. This record contains sediments for the past 3,000 years from which we will determine: 1) pre-settlement climate fluctuations; 2) timing of forest clearance; 3) chronology of crop introductions; and 4) climate changes during the last 1,000 years.

Dr. Richard Rainier, geo-archaeologist, mapped the entire Polie ditch. Using a Global Positionary System, Dr. Rainier was able to plot more than 2,200 points. This is the most detailed archaeological mapping ever completed on Easter Island. With this detailed survey, we can begin to systematically excavate the Pokie ditch, probably next year.

Mr. Kevin Buckley, Superintendent of the Gateway National Recreation area, U.S. National Park Service, working with Jose Miguel Ramirez, Director of the Easter Island National Park (CONAF), will incorporate the information obtained by the scientific teams in the development of a management plan for Easter Island. Easter Island has been designated a "World Heritage Site". It is estimated that the numbers of tourists visiting the Island will exceed 50,000 by the year 2025 and will continue to increase. It is important that in order to maintain its cultural heritage and preserve its monuments that a strict management plan be completed and put into place. The proposed NPS plan is included in this book.

Towards this end, we hope to establish an ongoing arrangement between the Science Museum of Long Island and the Institute of Pacific Studies on Easter Island. The function for this coordinated union will enable us to bring scientists from the United States and elsewhere to work each year on Easter Island, and at the same time allow Easter Island students the opportunity to work with scientist and American students. Stipends can be provided for students in this program. This organization will permit us to prepare work long in advance and eliminate duplicating research projects.

The restoration of Ahu Hanga Piko in Hanga Rau, was started in 1997 by the Science Museum of Long Island expedition. The completion was accomplished with the final placing of the broken moai head upon its body this year, under the direction of Easter Island archaeologist Sergio Rapu. Expedition members, village residents, members of the Chilean Navy, the Easter Island National Park (CONAF), elders of the tribe and tourists all attended this historic event.

Robert Hemm, Marcello Mendez and Garry Warren flew for the first time on Easter Island, a two-man powered parachute plane especially developed for the expedition. Logging more than 36 hours they flew at low levels and slow speeds, photographing the landscape for archaeological sites and monuments that would not be visible from ground level. The last time such an aerial photo survey was conducted on Easter Island was in early 1970 when Jacque Cousteau's son did it with a fixed-wing netra light. Unfortunately that early-70's flight ended in a crash and little data collected about the island.

This book now brings together the results of the three expeditions, identifies new areas of research, and hopefully will continue to inspire aspiring scientists to revisit this amazing island to explore and demystify this timeless enigma of human history.

PART I

DOCUMENTING THE WORLD HERITAGE SIGNIFICANCE OF EASTER ISLAND

Several Moai never totally excavated without a keel. Uncontrolled tourist access has eroded a large portion of the moai in the foreground from foot traffic.

Chapter 1

A Cultural Icon: Scientific Exploration into the World's Environmental Problems in Microcosm

JOHN LORET
Director, The Science Museum of Long Island; Past President, The Explorers Club

1. EASTER ISLAND: AN ECOLOGICAL TRAGEDY?

The most intriguing mysteries of human history are those posed by vanished civilizations. Anyone who has seen the abandoned structures of the Maya, Machu Pichu, or Angkor is moved to ask the questions: "Why did the societies that constructed these structures disappear? What lessons can we learn from their experiences, and, who can say that we on the planet Earth will not succumb to the same fate?"

Among all such vanished civilizations, the Polynesian society of Easter Island remains most outstanding in mystery and isolation. Easter Island is perhaps the most detached, fascinating, and intellectually intriguing locale on Earth. On this land mass of sixty-four square miles can be found material remains of a complex culture, which advanced to a high degree that produced gigantic stone statues weighing in the score of tons, a profusion of petrographs, large stone platforms, a written language, a systematic study of solar and stellar movements, and many other unique features. Yet in just a few centuries, the people of Easter Island consumed their forest, drove their plants and animals to extinction, and observed their complex society revert to chaos and cannibalism. (Flenley, J.R., 1993)

Easter Island, Edited by John Loret and John T. Tanacredi
Kluwer Academic/Plenum Publishers, New York, 2003

Easter Island perhaps is the most isolated inhabited land in the world. It is situated in the southern Pacific Ocean more than 2000 miles west of South America, 1,400 miles from the nearest inhabitable island, Pitcairn, of *The Mutiny of the Bounty* fame. Easter Island at latitude 27 degrees South and 110 degrees West has a mild climate, and its volcanic origin provides a fertile soil. In fact, Easter Island should be a miniature fertile paradise.

After the Polynesians on Easter Sunday, April 5, 1722 the Dutch explorer, Jacob Roggeveen, rediscovered the island. Roggeveen described the island as a wasteland, seeing nothing but withered grass, scorched and burnt vegetation, and complete poverty and barrenness. He did not see a single tree or bush over ten feet high. Recent botanical studies indicate that there are currently only 47 species of higher plants native to Easter Island, most of them grasses, sedges and ferns. With such a limited flora, the islanders Roggeveen encountered had no source for real firewood to warm themselves during the cold, wet, windy winters. Their native animals included nothing larger than insects. There was not even a single species of native bat, land bird, land snail, or lizard. The chicken was the only domestic animal that they had. (Roggeveen, J., 1908)

Early European visitors estimated the island's human population to be about 2,000. When Captain James Cook visited Easter Island in 1774, he concluded that they were Polynesians. Cook had a Tahitian with him who could speak with the Easter Islanders. Polynesians throughout Oceania have well-deserved fame as a seafaring people. Yet the Islanders, who came to Roggeveen and Cook's ships, did so by swimming or in small poorly constructed canoes, no more than ten feet long carrying one or two people. Cook and Roggeveen describe these vessels as being assembled with small planks, sewn together with twisted threads with no caulking between the seams. The boats leaked constantly and had to be bailed continually. Only three or four canoes were observed on the entire island.

With these flimsy crafts, Polynesians could not have sailed to Easter Island from even the nearest island nor could they engage in any type of offshore fishing. Since Roggeveen's rediscovery of Easter Island, investigators have not found any trace of the islanders having outside contacts. No rock or artifact from Easter Island had turned up anywhere, nor had anything been found on Easter Island that could have been brought by anyone other than the original settlers or by visitors since Roggeveen. (Cook. J., 1777; Roggeveen J., 1902) How did the islanders' ancestors reach Easter Island?

There is a strong logic on the side of Easter Island's initial discovery by Polynesians from islands some 2,000 miles west (perhaps Tonga or the Marguesas). Most of the colonization of the Greater Polynesian Islands, covering an area larger than the United States, took place well before Columbus, (or the Vikings) sailed the Atlantic and discovered America.

Accomplished Polynesian sailor navigators, along with their families, more than likely left their home, not to explore, or to trade, conquer or pillage, but more probably due to populations and resource pressures. Much like early American settlers seeking land who migrated west in the 1800's. Instead of Conestoga wagons crossing inhospitable deserts, mountains and rivers, the Polynesians sailed long stretches of oceans not yet explored. Like the western U.S. migrations, families, along with women and children, banded together and took with them all necessities including animals, plants and seeds, tools, and their culture and religion. They were committed for the long haul.

It is also very likely that in mid to late 500 A.D., just such a group of Polynesian immigrants under the leadership of Hotu Matua, set sail in a large double hulled wooden boat with 40 to 50 men, women and children of one or more families or clans, to set up a colony on a nearby island. They were most likely blown off course by a storm and found themselves on a long journey of some one and one-half to two months until they at long last sited land. It turned out to be Easter Island. Although they found Easter Island hospitable with adequate food, water and shelter, they must have eventually yearned for contact with others. Expeditions were sent out and returned without contact or didn't return. How many we'll never know.

The most striking features on Easter Island are its huge stone statues called "Moai"; over 200 that stood on large stone platforms, or "Ahu's," facing inland along the coast. Some of these stone slabs used in construction of the ahus weigh over 10 tons. At least 800 statues have been found in the process of being moved along ancient roads or in various states of completion in the volcanic quarry, Rano Raraku. The rock used in making the statues is volcanic tuff. Some statues weighing as much as 50 tons were moved as far as 8 miles. The largest statue, which stood on an Ahu, weighs over 90 tons, and an unfinished statue in the quarry is 65 feet high and weighs over 250 tons. The statues and slabs were cut with stone adzes made of hard basalt. The utter size and number of these stone monuments suggest that the population had to have been greater than 2,000. The densities of archaeological sites indicate a population greater than 7,000; some archaeologists estimate it could have been up to 20,000. (Sergio Rapu, 1999, personal communication).

Where did the first settlers come from? Dr. Thor Heyerdahl's raft and reed vessel voyages aimed to prove the feasibility of prehistoric contacts. He suggested that Polynesia was settled by advanced societies of American Indians, who in turn received their civilization from across the Atlantic Ocean from older societies of the Old World. Recent archaeological

Figure 1. Extinct Caldera at Rano Raraku

evidence indicates that early migrations came from the West and derived from Asia.

What happened to these early settlers? For the answer to this important question we must turn to other science disciplines, mainly archaeology, seed and pollen analysis, paleontology, and more recently, climatology. Modern archaeology started with the work of Dr. Thor Heyerdahl on an expedition in 1955-56. The earliest radiocarbon date of human activity was A.D. 400-600. The period of Ahu building started shortly thereafter. The period of statue construction came later and peaked about 1200 to 1500. Estimates of how the statues were carved and moved were obtained by employing twenty islanders using stone adzes. They could have carved a large statue in one year. Given enough timber and rope, 150 to 200 people could have loaded the statues on wooden sledges and dragged them overland on rollers. Using long logs as levers, statues could have been lifted on the Ahus. Ropes could have been made from the fiber of a small native tree called the hauhau. Only a small specimen the size of a bush was still alive on Easter Island when the Heyerdahl expedition visited in 1955. Hauling one statue would have required hundreds of yards of ropes. The barren landscape of Easter Island must have once been a sizable forest. (Heyerdahl, T., 1961) Pollen analysis obtained from coring a column of sediments from swamps in extinct volcanic craters (Fig. 1) contain both seeds and pollen of plants, the more recent record being in the upper sediments and the ancient record in the bottom deposits. The actual age of each layer can be dated by radiocarbon techniques. The pollen grains are then extracted from each layer and specific species of plants are identified microscopically. John Flenley of Massy University, New Zealand and Sarah King of the University of Hull, England completed the most comprehensive pollen work done on Easter Island to date. The data of Flenley and King revealed that Easter Island was not a wasteland at all. Dating back 30,000 years, long before human contact, Easter Island had a lush subtropical forest consisting of trees, woody shrubs, ground cover, herbs, ferns, and mosses. In the forest, grew tree daisies, hauhau trees, and the hardwood, toromiro tree. The most common tree in the forest was a large palm, *Jubaea chilensis*, commonly known as the wine palm and no longer found on the island, but still found in Chile. The wine palm can grow to over 80 feet with a 6-foot diameter. Logs from this wine palm could have been used to make large canoes or as rollers to move statues. The wine palm was also important as a food source yielding eatable nuts and a sap-like syrup or honey. (Flenley, J.R., 1993)

In April 1996, 41 years after Heyerdahl's Aku Aku expedition to Easter Island in which I took part as a crewmember and diver, Robert Hemm and I carried the Explorers Club flag #123 to Easter Island, which Heyerdahl took on his historic trip in 1956. Heyerdahl also took flag #123 on the Kon Tiki expedition in 1947. Flag #123 was also carried for the Ra I, Ra II, and Tigres expeditions.

Our purpose in 1996 was to obtain coral cores to study the phenomenon of past El Nino occurrences in the eastern Pacific. Coral has growth rings similar to those of trees that can be used to date past events (dendrochronology) and thus provide us with important clues about the past climatological conditions such as yearly water temperature, salinity and approximate rainfall. These cores can also tell us about climatological changes brought by El Niño in past times upon such island groups thus characterizing with more certainty the paleo-climate of these archipelagos. Finally, by comparing this data against the archaeological record, it might perhaps explain some of the earlier as well as the later cultural events that occurred on Easter Island and elsewhere in eastern Polynesia and during which years it was most favorable to travel eastwards from Central Polynesia.

Chronologies of coral cores are based on annual density bands revealed on positive points from X-ray image takes. Coral dates are reconstructed by counting the annual skeletal density bands using the date of collection as a starting point. Thin sections are then prepared for isotopic analysis by vacuum roasting for one (1) hour at 275 degrees C. Resulting gases are analyzed for O_{18} and C_{13} with a mass spectrometer. From this survey, the salinity and temperature of the sea at the time where the section was taken can be ascertained. (Beck, W. *et al.*, 1992)

2. WHAT IS EL NINO?

Over most of the world, climate alternates between summer and winter. Even in the tropics where the weather is warm year round, rainy seasons sometimes called monsoons, swing with the dry seasons. Each has its special pattern of prevailing winds.

Over time the human race has learned to adapt to changing seasons. Man has sown and harvested crops, bred livestock, launched fishing vessels, and planned hunting expeditions according to defined calendar dates. Tradition has influenced the way we schedule events and activities in construction projects, military campaigns, school closings, on-and-off seasonal rates for hotels, etc. However, the rhythm of seasons cannot always be relied upon. At times the tropical Pacific Ocean and other large areas of the global atmosphere do not conform to the regular patters, disrupting the lives of countless species of plants and animals along with hundreds of millions of humans. Scientists have constantly been working to understand these unusual competing rhythms that have been given the name of El Niño.

In normal years along the coast of Peru, the winds blow along the equator from east to west affecting the properties of seawater. Moving the surface water westward, colder, nutrient-rich water rises upwards to replace it, a phenomenon known as upwelling. This deep nutrient-rich water mixes with the surface water and in the presence of sunlight causes an eruption of microscopic plants, phytoplankton, which use the nutrients brought up with the deep cold water. This phytoplankton production, in turn, affects the lives of tiny sea animals called zooplankton, which graze on the phytoplankton and, ultimately, these are fed upon by creatures higher in the food chain, such as squid, fish, etc.

El Niño, the Spanish word for "the Christ Child", was first used by fishermen along the coasts of Ecuador and Peru to refer to a warm ocean current that typically effects the coast around Christmas and lasts for several months, and sometimes much longer. Fish are less abundant and most fisheries are closed down.

Over the past fifty years, nine El Niños have affected the South American coast. The weaker events raised sea temperatures 1 – 2 degrees C, but during the stronger episodes, such as occurred in 1982-83, the sea temperature was raised 6 – 8 degrees C and extended across the equatorial Pacific for over 5,000 miles. This event not only left its imprint on the marine life, but also altered climatic conditions around the globe. Within weeks, the ocean responded. Changes in wind speed and direction were recorded throughout the tropical Pacific. Sea level in the mid-Pacific rose 8 to 10 centimeters eastward to Ecuador, where the sea rose to over 20 centimeters. As sea levels rose in the eastern Pacific, it simultaneously dropped in the western Pacific, exposing and destroying the upper layers of coral reefs. Where sea levels rose, sea birds abandoned their young and scattered over a wide expanse of ocean in search of food. By the time conditions returned to normal along the coast of Peru and Ecuador, 25% of the year's fur seal and sea lion adults and all pups had died due to the lack of food sources. In addition to the disruption of marine life, droughts or torrential rains occurred in various parts of the world. Many scientists consider it the greatest disturbance of the ocean and atmosphere in recorded history.

Along with the changes in sea level, the barometric pressure also changes. When barometric pressure rises in the eastern Pacific, it usually falls in the western portion, and vice versa. This east-west season of El Niño has been termed the El Niño Southern Oscillation (ENSO).

A coring gun that was developed obtained a continuous core up to four feet in length using a 15 inch laser welded diamond drill bit manufactured by McMaster-Carr Company of New Brunswick, New Jersey. An adapter was machined to fit the coring bit to the "Sioux" air motor gun, model #1454. With air couplings, the air tube from the gun was fitted to the second stage of a SCUBA regulator, which used a standard scuba tan (see photograph). Three threaded extension rods, each one-foot long, were machined. With the core drill, a one-foot core could be obtained and then extruded into a PVC pipe, then an extension could be added to obtain a deeper core from the same hole. This was continued until the maximum depth of four feet was reached.

On April 9, 1996 divers Robert Hemm (MR 89) and Michel Garcia, archaeologist Claudio Cristino, owner of the vessel Jeff Lorton and I, as a marine biologist, set out to dive on a coral head near Ahu Akapu, off Hanga Roa. The coordinates with our global positioning systems (GPS) were 27 degrees, 08 minutes, 18 seconds South, 109 degrees, 25 minutes and 05 seconds West. The coral head was in 70 feet of water—an excellent test for our new coring drill. All went without difficulty. In fact, the entire coring did not take more than one hour. Our first four feet of coral had been obtained and would be sent to the Paleoclimatology Laboratory in Boulder, Colorado for analysis. In August 1997, we returned with a team of scientists and crew to obtain wider cores using a standard three-inch coral-coring device operated from a surface compressor.

3. WHAT DID EASTER ISLANDERS EAT?

Recent studies of garbage heaps (middens) on Easter Island indicate that the earliest settlers had the capability of fishing offshore. Throughout Polynesia archaeologists have found that 90 percent of the middens consist of fish or shellfish. On Easter Island, however, from the period of 900-1300 A.D., one third of the bones found were of porpoises. Yet nowhere else in Polynesia do the bones of porpoises account for more than one percent. These marine mammals had to be hunted by harpooning far offshore indicating that their craft had to be sufficiently seaworthy and constructed from large trees. Also bones found in the middens were of birds of over twenty-five nesting species which no longer can be found on the island, such as albatross, boobies, frigate birds, fulmars, petrels, prions, shearwaters, terns, tropic birds, barn owls, herons, parrots and rails. The bones of the Polynesian rat that the early settlers took with them were also well represented in the middens. In fact, rat bones outnumbered the fish bones. (Kuschel, 1963)

The early Polynesian settlers that first arrived on Easter Island some 1600 years ago found themselves in a pristine environment. Yet within a few centuries they completely consumed their forest. Preliminary pollen core studies indicate that in only a few hundred years after the first colonists arrived, Easter Island's forests were being destroyed. Charcoal from wood fires filled the sediments, while pollen of palms, trees, and woody shrubs decreased and even disappeared. The pollen of the grasses that replaced the forests became more abundant. By 1500 A.D. the palms became extinct. The hauhau tree did not become completely extinct. However, their numbers became so few that ropes likely could no longer be made from them. (Flenley, J.R., 1998)

The destruction of the island's animals was just as extreme as that of the forest. Every species of land bird became extinct. The coastal shellfish were over harvested. To make up for the loss of these food items, the islanders intensified the production of chickens that still could not meet the demand. Then they had to turn to their next abundant meat source available, humans, and whose bones are commonly found in the middens. With no wood left to cook, the islanders used sugar cane scraps, grass and sedges as fuel.

The evidence portrays a grim story of the Easter Islander society's decline and fall. The first settlers found a land with fertile soil, streams, abundant food, plenty of forest for building materials and a comfortable living space. They did well and multiplied. They begin building platforms, Ahus. Then, they carved and erected statues, Moais. As the years passed, the Moais and Ahus became larger and larger, many with red scoria top crowns weighing as much as ten tons. Was this a competition of "one-upmanship"? Did rival tribes try to outdo one another as an indication of wealth and power?

As the population increased the forest was cut more rapidly than it regenerated. The people used the land for gardens and the trees for canoes, houses, and fuel and to move statues. As the forest disappeared, they ran out of timber and life became more uncomfortable. Without trees the soil eroded by wind and rain, the nutrients were leeched out and the streams dried up. Without canoes, fishing declined. Intensified chicken production and cannibalism replaced only some of the lost resources. Woodcarvings of people with sunken cheeks and exposed ribs called Moai Kava Kavas suggest that the people were starving. With the shortage of food, local chaos ensued and a warrior society replaced the old centralized form of government. Stone points, Matoas that are made from volcanic glass (obsidian) were used on clubs and spears by warriors to kill enemies. For protection, people hid and lived in caves. On or about 1770, rival clans started to drop each other's statues, breaking the heads off. By 1884, the last statue was pushed off its Ahu platform. (Flenley, J.R., 1993)

As we look at Easter Island today, we can see a parallel with our own human society. Population is growing at a more rapid rate. We are exhausting our natural resources, our fisheries, forests and fossil fuels. We are polluting our air and drinking water and exposing ourselves to harmful UV-radiation by depleting the ozone layer. People are starving in many lesser-developed nations where central governments are yielding to gangs of thugs. Each year we find that we have more people and few resources to support them. If only a few thousand Easter Islanders with only stone tools managed to destroy their society, how can billions of humankind with steel tools, machines and atomic power fail to do worse? We have the experience, however, of knowing historically how other civilizations failed. From their legacies hopefully future generations equipped with sophisticated science and technological knowledge will find alternatives that were not available to the Easter Islanders.

REFERENCES

Beck J., Warren, R., Lawrence Edwards, Emi Ito, Frederick W. Taylor, Jacques Recy, Francis Rougerie, Pascale Joannot, Christian Henim. Sea – Surface Temperature from Carol Skeletal Strontium/calculi's Ratios, *Science* – 31 July 1992, Vol. 257, pp. 585-716.

Cook J., 1777, A Voyage Towards the South Pole and Around the World. Performed in His Majesty's ships "Resolution" and "Adventure". 2 Vols. London-Strahn and Cadell

Flenley, John R., 1993, The Paleocology of Easter Island and Its Ecological Disaster *Easter Island Studies* pp. 26-45.

Heyerdahl, T., and E.N. Ferdon (eds.), *Reports on the Norwegian Archaeological*

Expedition to Easter Island and the East Pacific – Vol. I Archeology of Easter Island, Monographs of the School of American Research and the Museum of New Mexico 24(1) London.

Kushel, G., 1963, Composition and Relationship of The Terrestrial Faunas of Easter, Juan Fernandez, Desventuradas and Galapagos Islands – *Occasional Papers of the California Academy of Science (San Francisco)* 44: 79-95.

Loret, John, El Nino's Wrath, *Explorers Journal*, Vol. 75 #1, Spring 1997, pp. 22-25

Rapu, Sergio, 1998 Personal Communication

Roggeveen, J., 1908, An extract from Mijnheer Jacob Roggeveen's official log of his discovery and visit to Easter Island in 1722.

Chapter 2

Rapa Nui National Park, Easter Island, Chile: An Eco-Tourism Outline with Issues and Suggestions

J.T. TANACREDI[1], K. BUCKLEY[2*], T. SAVAGE[2], and B. CLIVER[2]
[1]*Dowling College, Department Earth and Marine Sciences;* [2]*National Park Service;* [*]*retired*

1. INTRODUCTION

The goal of this suggested eco-tourism plan for Rapa Nui National Park in Chile is to guide the development of an environmentally sensitive plan to carry the park, a World Heritage site, into the future. A plan developed from an expanded version of this plan will guide the use and development of this world heritage site while simultaneously preserving and protecting and, where possible, enhancing the park resources. It must also take into account the three thousand residents of Easter Island, especially the indigenous Rapa Nui who have taken care of these resources for centuries and depend on them for spiritual and economic well being.

2. CULTURAL RESOURCE ISSUES

There are many issues having to do with protection and long term preservation of resources on Easter Island. Resource issues of major concern include:

Moai Issue – The most visible cultural resource to the world at large and the one that makes Easter Island one of a few visual icons of its type in the world. These hauntingly mysterious sculptures of ancient Rapa Nui are at significant risk not only from environmental deterioration, but potentially

Easter Island, Edited by John Loret and John T. Tanacredi
Kluwer Academic/Plenum Publishers, New York, 2003

from being touched and damaged by tourists. There are an estimated eight hundred Moai on the island and all are currently accessible to anyone at anytime.

Suggestion – All Moai should be protected from human impact by formally designating areas of viewing which do not allow visitors to come into physical contact with the Moai. Further research on ways of mitigating environmental deterioration (some have already been used experimentally) should be explored and implemented where appropriate.

Of the eight hundred existing Moai, only the best, most representative and well sited and protected should be accessible to visitors. Tongariki, in Hotuiti Bay is an example of this level of quality, protection and accessibility.

Figure 1. All Moai are subject to weathering and this statue near Hanga Roa is easily accessible by tourists. Photo by John T. Tanacredi.

Petroglyph issue – Petroglyphs or rock art, are present at many sites across the island. Again there is environmental deterioration recorded at these sites. Significant damage is occurring from visitors and "informal guides" who mark and walk on the resource. Livestock (horses, cattle) also walk on the exposed art and cause significant damage over time.

Suggestion – As with the Moai above, formal viewing areas, which are delineated in combination with education is called for. Enclosures can keep cattle and horses away from prime resources.

Historic House issue – Remains of all of the many types of historic houses on Easter Island are scattered across the island. Remaining stone foundations of various shapes are vulnerable to being walked on by ubiquitous resident cattle and horses. They are also not protected from visitors who might not understand their significance and who might cause unintended damage.

Suggestion – Keep cattle and horses away from vulnerable resources through the use of fences and cattle guards and other enclosures where appropriate. The island populations of these animals, which for the most part roam free, pose a significant ongoing threat to the future of Easter Island as a tourist destination. As with the Moai sites the best, most well protected, culturally significant, and accessible house sites should be presented and interpreted to visitors. The most obvious of these would be Orongo and Rano Kau. The rest should be protected from random visitation and possible destruction.

Lifeway issues – The art, language, customs and traditions of the Rapa Nui culture is real and is valued by all. It exists today essentially as it has for hundreds of years. Few cultures around the world, with such distinctive ancestral icons, have intact artifacts and traditions, which have changed so little through the centuries.

Suggestion – the lifeways of the Rapa Nui must be perpetuated to provide a complete picture of this most uniquely appealing culture. The existing cultural institute on the island should be expanded and made a center for education for and about the Rapa Nui for both islanders and visitors. The ability to travel to a place where a culture, it's lifeways and icons are intact and are reflective of centuries of habitation, yet holding great mystery, is unique to few places on this planet. This must be protected and interpreted for all. Only by enhancing, protecting and interpreting, in a balanced way, this very special culture, can its uniqueness be perpetuated and shared with a world which reveres the holistic integrity of this marvelous place.

3. NATURAL RESOURCE ISSUES

The natural resources of Easter Island, to a great extent, are open to people, animals and the vehicles of islanders. They comprise both seashore, open grasslands and forest with varying types of vegetation from low shrubs to the Eucalyptus plantations planted during the last half of the twentieth century. Little vegetation is indigenous and there are no native trees left on the island.

Forestation issue – As mentioned, Eucalyptus trees have been planted in extensive plantations over the last thirty to forty years to replace indigenous palm trees and other species that have been eradicated by Easter Island residents. Some efforts have been made to replace extirpated Palm species with other palms with limited success. Some indigenous species of other plants remain on the island but are threatened. A greenhouse has been built and is used to grow native species.

Suggestion – The need to reintroduce native plant materials which, need little or no care except to become established, can only have positive benefits for islanders and island animals by providing them habitat. The aesthetic appeal of plant materials on a fairly barren landscape will also provide visual interest and if well enough established in large enough stands could provide limited materials for construction as well as material for tourist related objects for sale. To the extent possible non-native plant materials should be avoided.

Seashore and Grassland issue – The seashore gives way to upland grasslands, which then rise to a few prominent volcanic peaks on the island. All of these areas are open to island cattle and horses and to vehicles (including Rano Raraku and Orongo). This accessibility causes soil erosion, denuded vegetation, litter, and pollution of freshwater bodies.

Suggestion – Formally designate use areas within Rapa Nui National Park so as to prevent indiscriminate access to areas where people tend to congregate. Designate parking areas, pathways and other people and animal management controls so that resources are not trampled or inappropriately used. Develop propagation areas, including a possible expansion of the greenhouse, for native plant materials and identify a program to reintroduce extirpated species where possible. This would be an ongoing program, island-wide and would take many years to implement. A healthy vegetative system would provide habitat for many island animals both existing and reintroduced.

Marine life issue – Fish populations around the island are being denigrated and continual pressure may cause fisheries to collapse totally. The scientific literature is depauperate regarding the biodiversity of this isolated island. New species have been identified in tidal pools alone, so that a more detailed inventory of marine species portends to increase species abundance. The NPS along with the AMNH established over 4,000 voucher specimens catalogued and maintained at the American Museum of Natural History in New York City.

Suggestion – In addition to possible commercial fishing restrictions, consider the development of land based aquaculture as a means of supplementing indigenous fish populations especially in light of increasing tourism to the island, which will place even a greater demand on existing populations. This potentially would be a business opportunity for islanders. Interpret the diversity of marine organisms of the island possibly using the marine environment as a scuba-diving eco-destination.

4. UTILITIES ISSUES

Power issues – Currently power to the island is provided by diesel generators whose fuel is brought in periodically from the mainland. This has the potential for toxic spills and causes air and noise pollution in the immediate vicinity. A failed delivery or major spill could cause the island to be without power for a period of time.

Suggestion – Consider the development of alternative energy systems for electricity generation including active solar systems (photo-voltaics), wind chargers or hybrid systems which would use both solar and wind. The solar access and continuous winds on Easter Island make these ideal alternatives for energy generation, they are renewable and non-polluting. These systems are now used very successfully all over the world and are competitively priced.

Water issues – Water is pumped from wells located across the island for domestic use including watering of cattle and horses. With a growing permanent and nonpermanent human and domestic animal population, the potentially significant fresh water draw down could result in serious water shortages. Additionally, pollution from failing septic systems could negatively impact water quality.

Suggestion – A second source of water should be developed. Given the amount of island rainfall and its variability, the government should develop ways of capturing rainfall runoff and storing it in cisterns. This would supplement the existing vulnerable aquifer and be a dependable source of fresh water during times of need and into the future.

Sewage issues – Septic tanks and leach fields are currently the predominant method of sewage treatment on the island. Over time the likelihood of failure of these systems is assured unless they are continuously maintained.

Suggestion – Alternative sewage treatment systems should be used to help prevent pollution of the aquifer. One of the most efficient and natural systems currently being employed is sewage treatment using constructed wetlands, which requires very low maintenance and has a lower initial cost of installation than other sewage treatment systems. The effluent flowing from these natural looking wetlands is tertiary treated and can be used for irrigation or allowed to soak into the ground. An experimental wetland could be developed and monitored.

5. TRAVEL AND TRANSPORTATION ISSUES

Transit issues – Currently there is no organized transportation system on the island. Visitors to the Rapa Nui National Park go independently to the various sites generally by automobile but also by horseback or by walking. Bicycles and motorbikes are also used. This independent access requires redundant ways of serving fluctuating populations of both permanent and non-permanent people. No map of the Parks' boundaries is available to casual visitors. During peak visitation periods (Dec. thru Feb.) little direction or control of access occurs.

Suggestion – Consider providing vehicles that carry multiple passengers and are non-polluting. This could be through the use of electric vehicles, which could be recharged using the phototvoltaic and wind electricity generators discussed above. Another option is the new hybrid cars being marketed by some Japanese manufacturers. These vehicles have both electric and gasoline motors in them and has ranges of approximately 800 miles per tank of gasoline. All taxis on the island should be pollution free if possible.

Road Issues – Access to the park is currently via some paved but mostly dirt roads. Dirt roads require constant maintenance to prevent erosion and to insure their viability.

Suggestion – Island roads should be improved using hardening techniques (without paving) on existing dirt roads. This technique will require less maintenance and the surfaces would be pervious and allow water to penetrate. This will help keep the aquifer recharged and will result in less erosion of the roadways.

6. HIKING TRAILS ISSUES

Few formal hiking trails exist on the island.

Suggestion – Trails which circumnavigate the island and pass near major areas of interest should be developed. Accommodations from camping areas to "bed and breakfasts" type accommodations could be developed to support this activity and provide cottage industry opportunities to islanders.

Airport Traffic Issues – Currently three flights per week land on Easter Island bringing mostly tourists and some other cargo. This is proposed to increase to seven flights per week. This will clearly place an increased burden on the island infrastructure and resources and will cause more goods to be imported to the island thus exacerbating the need for more landing space.

Suggestion – A study to determine island carrying capacity should be undertaken to determine ultimately how many people at one time can be on Easter Island without "significant and sustained damage" to resources. This should include looking at landing space as a controlling factor.

Currently there are two thousand Rapa Nui in residence on the island joined by one thousand Chileans and other nationalities. As a temporary limit on tourism the number of tourists should be limited to no more than one thousand at any one time so that at no time would the native population be outnumbered by others. This will not only provide protection against resource degradation but will help preserve the culture and tradition of the native population.

This "temporary capacity" would be used as a guide for infrastructure development. As new technologies increase the ability to allow more individuals on the island, (up to a point) this number can be adjusted. Additional visitors to the island will increase the economic viability of agriculture, aquaculture and the development of small to medium size cottage industries. Industries that make gift items out of recycled glass and aluminum as is done in other eco-tourism resorts and tour guides who are needed to guide and educate visitors and protect resources are a couple of examples.

7. SUMMARY

Rapa Nui National Park and Easter Island are on the edge of significant new visitation as worldwide access becomes cheaper and more frequent. Now is the time to put efficient infrastructure, human and otherwise, into place to preserve the world heritage of Easter Island. This necessity will spawn new businesses focused on eco-tourism and resource preservation. It is critical to develop a comprehensive eco-tourism plan for Easter Island which takes into consideration all of the aforementioned and more. Unless this is done and implemented, the longevity of the resources of Easter Island

and Rapa Nui National Park will be in jeopardy. This will result in a collapse of the economic base and way of life of islanders. It is clearly in everyone's interest to treat these world-class resources with great sensitivity. The quality of life of islanders will be determined by the quality of care given the environment and the very special one-of-a kind resources that exist in that environment.

REFERENCES

Tanacredi, J.T. (1994), *Gateway Visitors Companion*, Stackpole Books, Inc., Mechanicsburg, PA, USA.

Chapter 3

Documenting Petroglyphs on Easter Island

BLAINE CLIVER
Manager, Historic American Buildings Survey/Historic American Engineering Record, National Park Service, Washington, D.C.

1. INTRODUCTION

Lost in the vastness of the south Pacific Ocean, Easter Island is the most isolated inhabited place on the face of the Earth. The first European to land on the island was the Dutch admiral, Jacob Roggeveen, who paid it a single day's visit on Easter Sunday in 1722. An expedition dispatched by the Spanish viceroy of Peru rediscovered the island in 1770, calling it San Carlos. To the original Polynesian settlers, it was simply "the land," the center of the world, Te Pito te Henua. The Spanish were the first to report that the aborigines had their own local form of writing and estimated the population to be some 3,000 persons. Civil war or plague seems to have raged on the island before the arrival of the British navigator, Captain James Cook, in 1774. He found a decimated, poverty-stricken, Polynesian population of only about 600 or 700 people, and observed that the large statues were no longer venerated, many having been overthrown.

Today Easter Island, commonly called Isla de Pascua in the Spanish of Chile, the country to which it is attached, or Rapa Nui as it is known in its natives' language, has been occupied by humans for only about sixteen-hundred years. Yet, during this short period of habitation it has had significant historical development and has produced some physical aspects of its culture that rival the monuments of many older and more established cultures.

Easter Island, Edited by John Loret and John T. Tanacredi
Kluwer Academic/Plenum Publishers, New York, 2003

The island's archaeological sites are in such abundance as to give the impression of a vast outdoor museum. Much research has been accomplished over the years and many new things are being found out about the people and their culture. However, despite many recent discoveries, the island's past lies under a shroud of mystery and controversy.

The island has great beauty with three large mountains rising to a maximum of 600 meters. In between, smaller volcanic cones create a rolling, hilly though rocky, landscape. Some cones have distinctive types of stone found on them, such as the brownish tuff of Rano Raraku, the famous statue quarry, or the red scoria of Puna Pau where the topknots (pukao) for statues were obtained. Several of the sites, particularly the awesome quarry at Rano Raraku and the beautiful ceremonial site of Orongo, are spectacular and, in their isolation, are different from anything found elsewhere in the world. Besides the famous statues (moai), recognizable from the worldwide publication of numerous photographs, there are thousands of less-well-known rock carvings or petroglyphs that permeate many areas of this marvelous island. It is these petroglyphs and the method of their documentation that is the focus of this paper.

2. THE FIRST EXPEDITION

In August of 1998, I was fortunate to be part of an expedition to Easter Island led by Dr. John Loret, Director of the Long Island Museum of Science. Having been a member of an earlier expedition to this island led by Thor Heyerdahl, John Loret brought with him a knowledge and love for this lonely place that was an encouraging and educational aspect of the trip. This first expedition consisted of groups of professionals from the fields of archaeology, biology, planning and documentation together with volunteers who assisted on the various teams. The team that I led, which included from 3 to 5 volunteers depending on the day and activity, had the objective of documenting rock art, or petroglyphs, through the use of photogrammetry; documenting a moai was also attempted.

The first question that might be asked is "why document or record these art objects?" and, if this question is answered satisfactorily, "how can it best be done?" We of course should properly document this rock art because it exists as a feature of a past culture that was once a part of humanity and now is under great pressure from environmental as well as human actions that threaten its preservation. Carved into basalt, or what for the most part is a volcanic tuff, the petroglyphs are subject to erosion from rain and internal pressure from soluble salts diminishing the depth of the base relief. Numerous cattle as well as horses roam the island at will, walking and excreting upon the carved surfaces of those petroglyphs lying on the surface of the ground. However, it also is important to record these works of art so that it may be possible, in the future, to see many of these works through the Internet or, in virtual reality, without exposing them to the rigors of tourism. And, when the erosion has taken its toll, there will be a record that may be measured and viewed orthographically, not seen at the disadvantage of an oblique angle or in perspective.

Given that these are cultural objects worthy of care and documentation, how can this recording best be done? This was the question that I had to consider from the vantage point of over 10,000 kilometers, not having had the opportunity of a previous visit to the island. Photogrammetry was an obvious consideration for two reasons: one, the photographs could be taken on the island and the processing done later when I had returned to my office, and two, with the proper program and computer equipment plus a day-load tank, the processing could be done on the island if the local staff wished to continue the work. Given the size of the objects to be recorded, most are one to three meters in size, it seemed that a 35mm. camera would be the most useful means of photometric recording, being easily portable with the possibility of local developing (use of a more sophisticated Metrica camera would have been extremely difficult because developing the 100mm. roll film used in this camera was impossible on Easter Island and would have required shipping to the mainland causing delay in processing the image. Also, it was felt that for the cost, the 35mm. camera would produce better resolution when the images were scanned than what could be produced by a comparable digital camera (the recent developments in digital photography have shown this not to be true and a digital camera was used quite successfully on the second visit). For software PhotoModeler 3.1[1] was selected because of its ability to work off the computer monitor without having to digitize points into the computer from a digitizing table (the shipping of large equipment to Easter Island was found to be difficult because of customs restrictions on the mainland of Chile).

3. WHAT IS PHOTOGRAMMETRY?

Shortly after the discovery of the Daguerre process of photography, the first stereograph or stereoptic views were produced. By the mid 1850's this process had been perfected and "stereographs," as negative images, were produced of buildings[2]. At the same time as the stereograph was being

[1] PhotoModeler 3.1 is a photogrammetry based software developed and distributed by PhotoModeler, and can be found at <http://www.photomodeler.com

[2] Darrah, William C., *The World of Stereographs*, p. 6.

perfected, developments were initiated in the art of photogrammetry. In 1849, Aimé Laussedat, an engineering officer in the French Army, began his efforts to prove the value of photographs primarily for the preparation of topographic maps. In 1859, after some unsuccessful attempts at aerial photography using kites and balloons, he announced to the French Academy of Science his use of a phototheodolite for surveying. In 1867, at the Paris Exposition, he exhibited this instrument, the first known of its kind and, in various versions, what was used for architectural photogrammetry until after the Second World War[3]. Although much of the early direction of photogrammetric development was toward topographic surveys, Laussedat did experiment with graphically generating a site plan of Chateau de Vincennes in 1850 (Fig. 1) from photographs using perspective sight lines drawn from numerous camera stations[4].

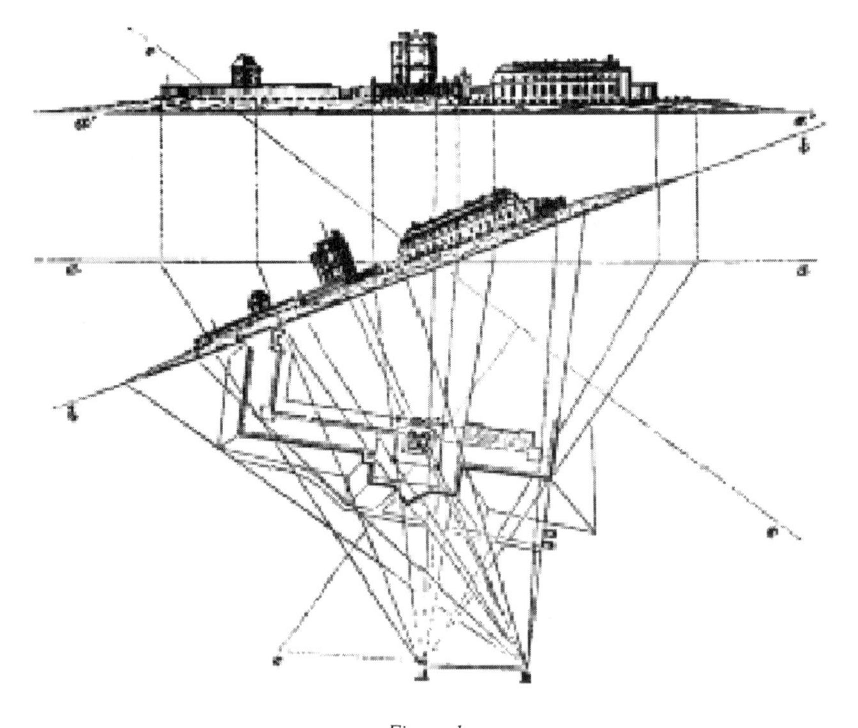

Figure 1.

[3] Whitmore, George D., The Development of Photogrammetry, *Manual of Photogrammetry*, p. 2-4.

[4] Carbonnell, Maurice, L'histoire et Situation Presente des Applications de la Photogrammetrie a l'architecture, *Application of Photogrammetry to Historic Monuments*, p. 3.

Laussedat's work was built upon by the German architect Albrecht Meydenbauer. Meydenbauer is credited with the first complete recording of an architectural monument in 1867[5]. Under his direction in 1885, multiple photographs of single buildings were collected. These photographs were from widely separated and carefully surveyed camera stations. With these photographs, and employing conventional graphic methods, architectural plans and elevations could be plotted. The process, as described in a late-nineteenth century surveying handbook, was as follows: "If a photograph be taken from a point whose position is already known, the direction of the axis of the object glass and the focal length of the lens being also known, and the line of the horizon being marked on the picture, then the picture can be laid down on a sheet of paper on which it is desired to plot the survey, and will give the direction from the point of observation of all the points in the picture whose position is required. Two photographs of the same objects taken from different known points define completely the position of each object, and also enable altitudes to be calculated or graphically determined. The method is exactly that of the plane table, the difference being that a great part of the work which with the plane table is done in the field, by the photographic method is done in the office."[6] Crucial to this methodology was the development of precise instruments. The invention in 1895 of the Bridges-Lee phototheodolite combined the compass direction and the angular measurement into the photograph making the graphic plotting easier and more precise[7].

At the beginning of this century the application of stereophotogrammetry, as opposed to graphically-projected analytical, or convergent, photogrammetry, for the production of architectural drawings was pioneered in Austria by Eduard Dolezal. As shown in Figure 2, "The process of taking stereopairs consists in setting up two tripods, leveling the phototheodolite, and photographing at each station at an angle perpendicular to the base line between the tripods [non-convergent]."[8] In spite of Dolezal's efforts, it appears however, that the use of stereophotogrammetry, including mechanical optical plotting, was not applied extensively to architectural recording until about the time of the Second World War--possibly owing to its cost as compared with the precision obtained. With the development of the airplane in the early-twentieth century, photogrammetry for map making became far easier and with this ease came more rapid development in the

[5] Koppe, R. & R. Meyer, Meydenbauer and His First Recordings (1878-1905), *Photogrammetric Surveys of Islamic Architecture*, p. 9.

[6] Whitelaw, John, *Surveying as Practised by Civil Engineers and Surveyors*, p. 112-114.

[7] Fryer, D. H., Cartography and Aids to Navigation, *A History of Technology*, Vol. V, p. 448.

[8] Borchers, Perry E., Architectural Photogrammetry, *The Journal of the A.I.A.*, June 1962

process as a whole, including better and more precise instruments for photographing and plotting.

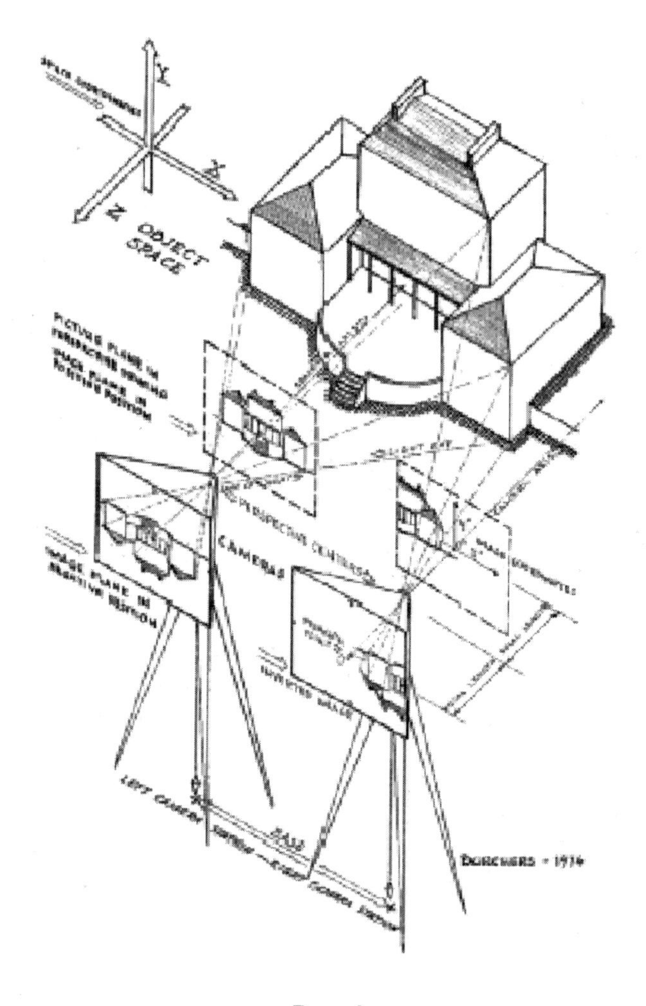

Figure 2.

From these initial efforts the use of photogrammetry proceeded to expand worldwide. In 1968, under the umbrella of ICOMOS, a symposium was held at Saint-Mande, outside of Paris, "...to define the requirements of the architect...[and] to discover the orientation to be given to photogrammetric methods as applied to...historic monuments."[9]. This symposium showed the extent of the use of photogrammetry at that time, especially in Eastern Europe, and the questions raised concerning its use in the recording of monuments. It was noted that accuracy was expensive and by accuracy it was meant a precision or tolerance of plus or minus 1 cm. Therefore, for economical considerations several levels of recording were suggested that would suffice for 1) diagrammatical purposes, 2) general impression or expressive drawings and 3) accurate recording. The first two could be obtained from rectified photographs and would use a precision of less than 1:100, but accurate recording required the use of stereopairs plotted with machines[10]. At this symposium there also was expressed a concern for establishing archives of photogrammetric plates and with this a questioning of the efficacy of such archives for the future considering the changes that might occur in the equipment employed. Interestingly, a thought was provided that by using electronic calculation to increase the speed of the process, the method of Meydenbauer might be reestablished and points could be plotted in space[11]. This in fact is one of the systems, using computerization, that today is applied to photogrammetric recording.

Figure 3.

[9] Summary Report on the Symposium held at Saint-Mande, France, 4-6 July 1968, *Application of Photogrammetry to Historic Monuments*, International Council of Monuments and Sites, Paris, 1969, p. 158.

[10] Paquet, Jean Pierre, Panorama des apports possibles de la photogrammétrie dans les différents champs d'études et de travaux de l'architecte. Possibilités et insuffisances., *Application of Photogrammetry to Historic Monuments*, p. 137-140.

[11] Summary Report on the Symposium, op. cit., p. 163.

By using photogrammetry it was planned to create an orthographic image of the petroglyph (Fig. 3) that could either be enhanced graphically to show the carving or, converted to a drawing either by hand or with a computer. The petroglyphs were carved onto the surface of stones that were, for the most part, flat but not entirely so. Most of the stone surfaces were rounded or in some manner irregular. Therefore it was important to characterize the surface topography in order that the final image be correct and to scale. The simplest way of doing this would have been to snap chalk lines in a grid pattern onto the surface of the stone—each photograph would have the same easily recognized points—thus the curvature could be characterized by the grid. However, using chalk would create marks not easily removed and detract considerably from the aesthetic quality of the petroglyph. Therefore, another approach was tried that consisted of using drafting tape to fasten white string in a grid pattern on the stone surface (later I was to find that a similar approach was used in hand-measuring the petroglyphs, shown in Georgia Lee's The Rock Art of Easter Island). The drafting tape was easily applied, left no residue on the stone, and the string could be reused. With these parameters decided upon, we landed on Easter Island on the evening of 13 August 1998. After a day of orientation, and with three volunteers, we set out to try our hand at recording the Rapa Nui petroglyphs. The first of the petroglyphs recorded were at Orongo, the site of the birdman cult.

4. ORONGO

A description of Orongo petroglyphs would be incomplete without a discussion of the birdman cult, the activities of which were associated closely with the site[12]. The basis of the ritual was obtaining the first egg of the season from the islet of Motu Nui, the place where great flocks of seabirds would come every September. Contestants were men of importance who probably selected servants (hopu) to represent them. The hopu descended Orongo's sheer cliffs and, using reed bundles for floats, swam to Motu Nui carrying provisions. When the hopu waiting on Motu Nui suspected that eggs were being laid, they came out of their hiding places. The hopu which procured the first egg would scoop the egg into a small reed basket tied to his neck, swim back, climb the cliff, and present the egg to his employer, or clansman, who would then become the new birdman. A celebration would begin with fires lit to inform the island that a new birdman reigned. The new birdman shaved his head, eyebrows and eye-lashes, and painted his head white. Taking the egg in his palm on a piece of tapa cloth,

[12] Drake, Alan, *Easter Island, The Ceremonial Center of Orongo*, p. 41.

he and his companions danced down the mountain to Rano Raraku, near the statue quarry, where the new birdman lived for the year under strict tapu. He was not allowed to eat or associate with his family nor was he to wash, shave or cut his nails for a full year until there was a new birdman. The following year a new birdman replaced the old one who then returned to ordinary life. However, a birdman was considered to be sacred throughout his lifetime and was given special honors at death[13].

Figure 4.

The petroglyphs are symbols of life. As Alan Drake says in his book, Easter Island, the Ceremonial Center of Orongo, "This new myth and cult of birdman which replaced Rapa Nui's traditional religious practices and beliefs, contains powerful symbolic elements of death and rebirth. These include the symbolism inherent in birds, eggs, figures in praying or fetal positions, descent into the great ocean, re-climbing the cliff with a sacred egg, spring renewal rites, the shaving of the head, taking on a new name, undergoing ordeal and confinement. The motif of birdman appears to be an archetypal symbol which arose from necessity out of the collective unconscious of the Rapanui, a response to extreme societal stress and the deeply felt needs of the island's population."[14] (Drake, A., 1992)

[13] Lee, Georgia, *An Uncommon Guide To Easter Island*, p. 44.
[14] Drake, op. cit., p. 31.

Orongo is at the southwestern end of the island on a high promontory formed by the partially collapsed wall of the volcanic crater Rano Kau. It was here at the sacred precinct known as Mata Ngarau that the ceremonies of the birdman cult took place. It is in association with this cult that the petroglyphs were created. Of the 1,785 petroglyphs at Orongo, 375 depict the birdman which is a stylized man in prayer having a bird's head[15]. The site of the recorded petroglyph (Fig. 5) is on the edge of a shear cliff falling off into the sea. Therefore, the location of camera stations was limited. The volunteers quickly learned the technique of taping the string onto the petroglyph surface in a 15 to 20cm. grid pattern, the dimensions of which were not critical since it was only the intersections that mattered (Fig. 6). The surface was then photographed from three station points, the strings were removed and the surface photographed again from near one of the previous camera positions so one photograph would be without the strings.

When the photographs were processed in the computer the grid supplied by the string established the topography of the stone surface and the image of the photograph without the string was imbedded into the grid. Because of the relatively small size of the petroglyph, the transfer of points from the photograph having the string grid to the photograph without the grid was quite simple since there were enough irregularities to the surface that placing the locations of the intersections of the strings could be done with adequate precision. This was especially true when the photograph without the string

Figure 5.

[15] Lee, op. cit., p. 69.

Figure 6.

grid duplicated one of the photographs with the grid. (Of course, one could use a photograph with the string grid in place if it was anticipated that the orthographic image only would be used to produce a later drawing from the image. Otherwise, the string grid detracts from the appearance of the image.)

After using the computer program to create a three-dimensional image of the stone surface, the image is saved as an orthographic file that can be opened in any graphic program such as PhotoShop. In the graphic program image files can be enhanced by adjusting the brightness and contrast. Increasing the contrast can give the image more of the appearance of a drawing (see Fig. 3). A hard copy of the image can be traced over to produce what many people would consider a more artistic rendition of the petroglyph than that produced by a computer, or the image can be traced in the computer on a different layer in a graphic program.

5. THE SECOND EXPEDITION

The following August a second expedition led by Dr. John T. Tanacredi, National Park Service, Research Ecologist (Retired) and presently Chairman of the Department of Earth and Marine Sciences at Dowling College in Oakdale, LI, New York, went back to Easter Island and, again, I was fortunate to be along. This time I brought a digital camera with me. The camera, a Kodak 265, proved to be an enhancement over the 35mm. camera used the previous year. The resolution of the Kodak 265 was equal to a scanned 35mm. image. In addition, digital images from the camera, taken during the day, were transferred that evening to a laptop and processed; immediate results could be seen.

On this trip the emphasis was recording petroglyphs at other sites on the island. At Paka Vaka are found numerous marine petroglyphs. An octopus and a tuna were selected for recording using the same technique of applying string as was used the previous trip at Orongo. The intersection points of the strings were entered into the computer from three photographs. Each point is given a number by the computer program and from this information the computer is able to locate each point in space through assigning measurements to three axes. With this information the surface curvature is determined. This allows for the accurate projection of the orthophoto or orthographic image of the rock surface. (Figs 7 & 8)

Figure 7.

A comparison of a normal (perspective) photograph of a tuna (bonito) petroglyph and an orthophoto of the same petroglyph shown in Figure 9 illustrates the difference in the resulting images. This same orthophoto is compared in Figure 10 to a hand-measured drawing of the same petroglyph taken from *The Rock Art of Easter Island*[16]. Clearly the orthophoto is dependant on lighting for its definition, whereas greater detail can be

[16] Lee, Georgia, *The Rock Art of Easter Island, Symbols of Power, Prayers to the Gods*, p. 79.

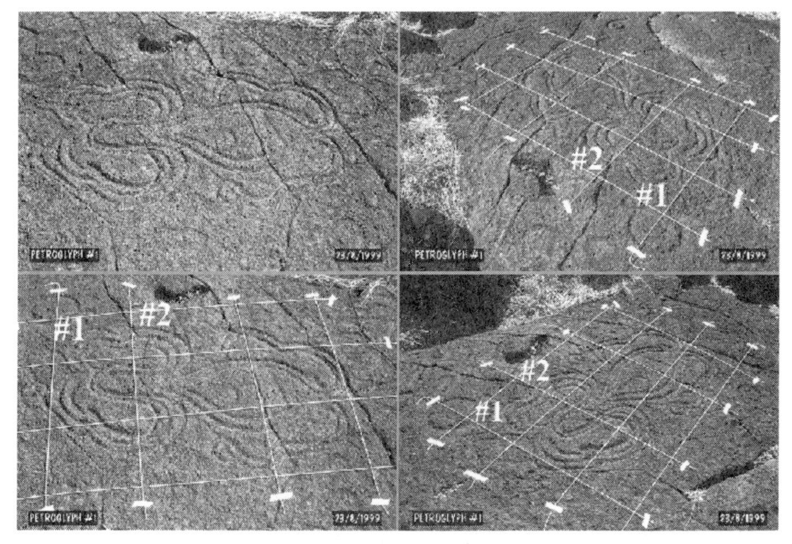

Figure 8.

Figure 9.

included in a drawing since the eye can discern elements not highlighted in the photograph. However, the time involved in creating the orthophoto has to be less than in the hand-measuring process, and the accuracy of line form greater. It is possible to combine the two approaches, once the orthophoto has been created, by returning to the site to note and add detail to the image that may not be found in the original photograph.

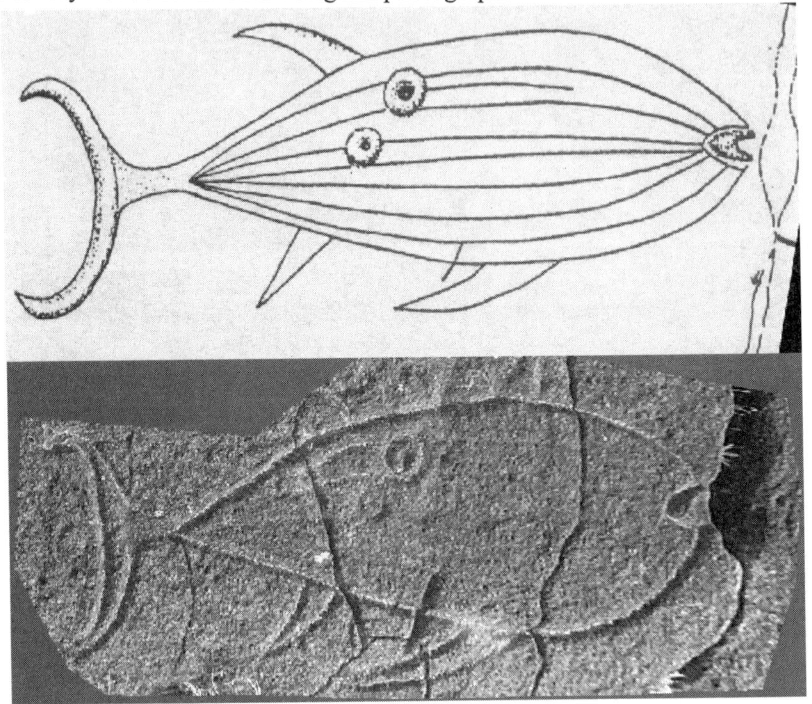

Figure 10.

6. CONCLUSION

It is possible to accurately record patterns cut into a rock surface through the use of a photographic system known as photogrammetry. The images thus created are without perspective and measurable in two dimensions. Being dependant on natural light for highlighting the subtle surface markings, it is not always possible to record small or faint details. Acquisition of these more obscure details can obtained through hand-measuring after the orthophoto has been created. It is estimated that for most sites the string grid can be applied by one or two persons to a petroglyph in less than 10 minutes since the grid does not have to be laid out with any specific measurements or precision. The photography can be done in a few minutes. Processing in the computer may take at the most 30 minutes depending on the number of points involved.

While on the island, experiments were made in recording the moai, or statues, of Easter Island. The statues, carved from tuff, a soft volcanic stone, range in height from 3 to 12 meters (10 to 40 feet), some weighing more than 50 tons. The same method of taping string onto the stone surface was used to create the surface topography in recording a moai. (A method of placing just drafting tape crosses on the surface was tried but was found to be more difficult when processing in the computer because of the difficulty in identifying individual crosses in each photograph.) The moai recorded with this method, two, were both lying face up, therefore allowing the documenting of the features which would have been impossible, owing to the weight, for those lying face down. (Without ladders, the standing moai would have been difficult to have taped string onto.) After plotting the points of string intersection and processing in the computer, wire frame images were obtained. Into these images the stone texture from the photographs was imbedded. However, because of the many surfaces, and irregularities, the value of this process to produce a measurable, or orthographic, image is questionable considering the time involved.

Although many of the petroglyphs are in the round or on several sides of a stone, it is the surface that is significant and what is important to record. The configuration of the surface is the art, and it is the carving on the surface that should be preserved. With the moai it is the shape and form not what is on the surface that is significant. Photogrammetry can be used to record the overall dimensions of moai, or the specific shape of surfaces for alignment when reattaching heads to bodies–many moai lie broken on the ground or on their stone platform (ahu). However, the recording of the moai, as with a statue, may best be done by other methods, quite possibly through the new and developing technique of laser scanning which is able to create a cloud of millions of points, each point located in space, in a matter of a few minutes. Through computer programs, the cloud of points can be converted into light and dark pixels creating a virtual image in three dimensions. As this technology becomes more portable the ease of using it to record moai on Easter Island will become greater. These moai are deteriorating each year and, if we are to retain any aspect of their appearance, we need to consider a serious approach to their documentation while they still retain the characteristics that make them unique.

There still remain hundreds, if not thousands, of petroglyphs undocumented on Easter Island. With the growing tourism, as well as the continuing threats from environmental erosion and human folly, the preservation of Easter Island's cultural resources may lie on the brink of disaster. In this situation it is feasible for photogrammetry to play an important role in recording some aspects of this island before much of these elements of a past culture are lost forever. It would not be an enormous effort, nor one requiring extensive time in the field, but to accomplish such a program will require determination and some support. To do nothing to document these resources in three-dimensions, with the methods now available, will be to sit by and watch them deteriorate at an ever increasing rate.

REFERENCES

Borchers, Perry E., The Measure of the Future and the Past, The Journal of the A.I.A., The American Institute of Architects, Washington, DC, October 1957,

Borchers, Perry E., Architectural Photogrammetry, The Journal of the A.I.A., The American Institute of Architects, Washington, DC, June 1962.

Borchers, Perry E., Photogrammetric Recording of Cultural Resources, National Park Service, Department of the Interior, Washington, DC, 1977.

Carbonnell, Maurice; L'histoire et Situation Presente des Applications de la Photogrammetrie a l'architecture, Application of Photogrammetry to Historic Monuments, International Council of Monuments and Sites, Paris, 1969.

Charola, A. Elena; Easter Island, The Heritage and its Conservation, World Monuments Fund, New York City, 1994.

Charola, A. Elena; Death of a Moai, Easter Island Occasional Paper 4, Easter Island Foundation, Los Osos, CA, 1997.

Darrah, William C.; The World of Stereographs, W. C. Darrah, Gettysburg, PA, 1977.

Drake, Alan; Easter Island, The Ceremonial Center of Orongo, Cloud Mountain Press, Old Bridge, NJ, 1992.

Fryer, D. H.; "Cartography and Aids to Navigation", A History of Technology, Charles Singer ed., Vol. V, Oxford University Press, New York and London, 1958.

Hanke, Klaus; Accuracy Study Project of Eos Systems' PhotoModeler, available on <http://www.photomodeler.com>, 1997

Koppe, R. & R. Meyer; Meydenbauer and His First Recordings (1878-1905), Photogrammetric Surveys of Islamic Architecture, International Council of Monuments and Sites, Tunis, 1988.

Lee, Georgia; An Uncommon Guide To Easter Island, International Resources, Arroyo Grande, CA, 1990.

Lee, Georgia; The Rock Art of Easter Island, Symbols of Power, Prayers to the Gods, University of California, Los Angeles, 1992.

Paquet, Jean Pierre; Panorama des apports possibles de la photogrammétrie dans les différents champs d'études et de travaux de l'architecte. Possibilités et insuffisances, Application of Photogrammetry to Historic Monuments, International Council of Monuments and Sites, Paris, 1969.

Whitelaw, John; Surveying as Practised by Civil Engineers and Surveyors, 8th ed., The Technical Press, London, 1936.

Whitmore, George D.; The Development of Photogrammetry, Manual of Photogrammetry, Chapter I, American Society of Photogrammetry, Washington, DC, 1952.

Chapter 4

The Corals and Coral Reefs of Easter Island - A Preliminary Look

DENNIS K. HUBBARD[1] and MICHEL GARCIA[2]
[1]*Department of Geology, Oberlin College, USA;* [2] *S.E.E.M. Orca, Ltd., Caleta de Hanga Roa, BP. 21, Isla de Pascua, Chile*

1. INTRODUCTION

Easter Island sits in geographic isolation in the southern Pacific Ocean at latitude 27° 8' S and longitude 109°20' W, nearly 4500 km west of Chile (Fig. 1). It is a small island near the western end of a chain of volcanoes that are related to hot-spot activity dating beyond 3.5 million years before present. The small island of Sala y Gomez to the east and a series of submerged platforms are an extension of a volcanic lineament along the edge of the Nazca Plate (Kruse, *et al.*, 1997; Newman and Foster, 1983).

The island is triangular in shape, with a major volcanic peak at each corner, Terevaka (>500 m) to the northwest, the Poike Peninsula (300 m) to the east, and Rano Kao (>400 m) to the south (Fig. 1). In addition to anchoring the island, these peaks have great cultural significance. The Poike Peninsula was the site of fierce fighting that decided the ruling class of the island. A smaller, subordinate cone at Rana Raraku was the main site of quarrying for the large ceremonial statues that are the best-known cultural remains on the island. These "moai", some of which exceed 10 meters in height, were carved by hand, apparently lowered by hemp-like ropes to the base of the volcanic cone and moved to a number of sites (ahu) all across the island. Their significance and their sudden abandonment as cultural icons remain as two of the most important anthropological mysteries surrounding the island's history.

Easter Island, Edited by John Loret and John T. Tanacredi
Kluwer Academic/Plenum Publishers, New York, 2003

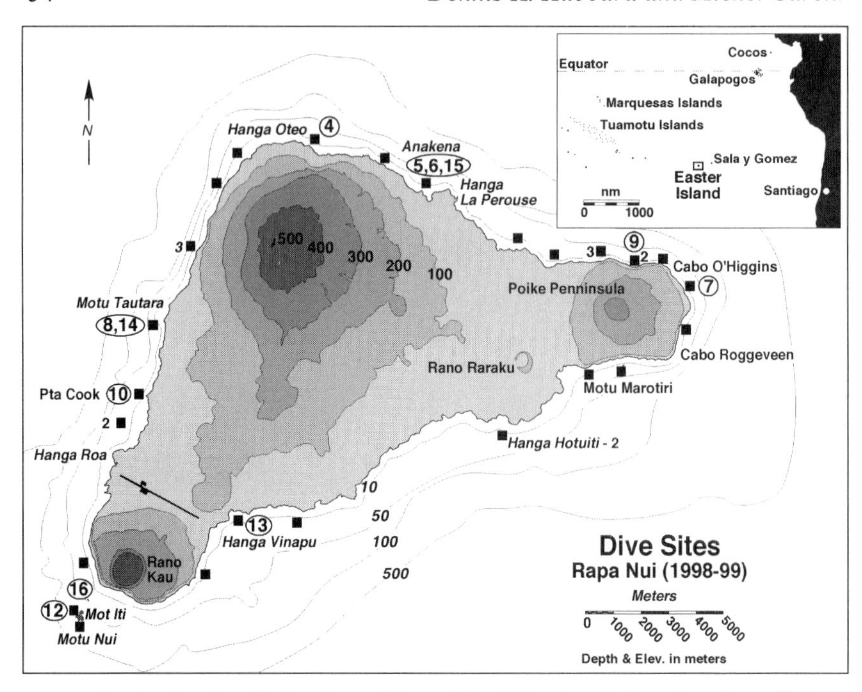

Figure 1. Map showing the location of Easter Island (inset) and dive sites along all three coasts. Dive locations are shown by black squares (the number of multiple dives is shown adjacent to the box). The locations of photographs included in later figures are also indicated (numbers in circles). Easter Island occurs in the eastern half of the clockwise gyre formed by the Peru Current, the South Equatorial Current, the East Australian Current and the South Pacific Current. Coral larvae picked up from the Indo-Pacific atolls to the west and Great Barrier Reef off Australia must normally pass through colder southern waters en route to Easter Island. However, larvae can reach Easter Island from the west during equatorial current reversals associated with El Niño.

Rano Kau was a later ceremonial site related to worship of Tangata Manu, the "birdman". As nesting season started, a warrior chosen from each clan presumably climbed down the precipitous volcanic cliffs and swam through strong swell and treacherous currents to Motu Nui, a small islet a kilometer off shore. He picked up a tern egg, swam back to the main island and climbed up the cliff. The first warrior to reach the top with an intact egg presented it to his chieftain, who became the leader of the island for the coming year. In this sense, politics appear to have remained unchanged across time and cultures.

2. METEOROLOGY/OCEANOGRAPHY

Easter Island sits in the eastern half of the counter-clockwise South Pacific gyre. This circulation pattern requires that surface waters normally pass through the colder, southern ocean en route to Easter Island, making the survival of coral larvae from the western Pacific unlikely. Oceanic currents are generally mild, although strong currents were observed during our field studies between Isla de Pascua and the small cays (motu) off its southern tip near Rano Kao. Summer water temperatures around Easter Island peak in March, typically approaching 25°C (Fig. 2). While winter temperatures average near 20°C in the months of August and September, the Chilean Hydrographic Institute has measured low temperatures approaching 17°C (DiSalvo, *et al.,* 1988). Randall and McCosker (1975) reported a low of 15.7°C in 1963. These temperatures are near or below the lower tolerance limit of scleractinean corals.

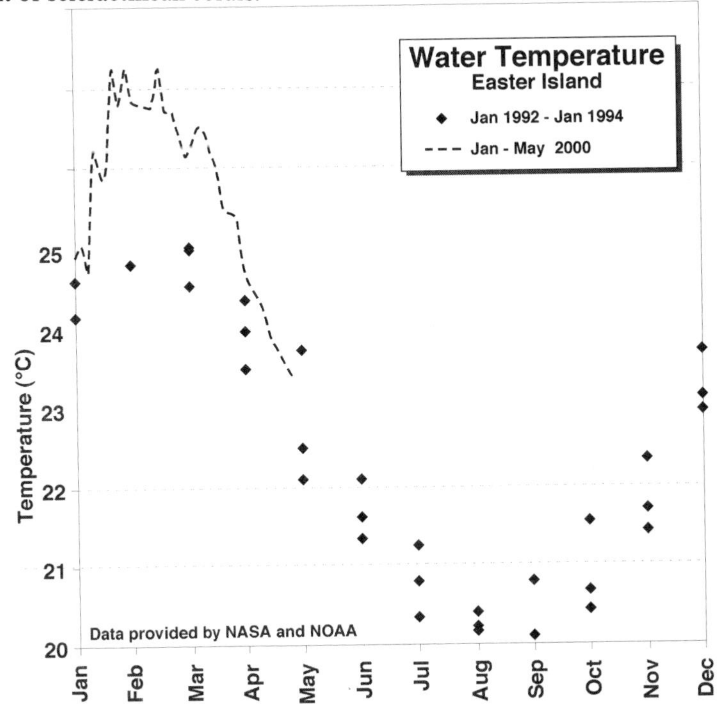

Figure 2. Monthly water-temperature measurements from Easter Island. Each point represents a monthly average computed from hourly data between February, 1992 and January, 1995. Elevated temperatures associated with La Niña in early 2000 are shown by the dashed line. Raw data were provided by the U.S. National Oceanic and Atmospheric Administration (Station 9962420).

In general, summer winds are moderate, averaging near 4 knots, with gusts averaging 10-12 knots (Fig. 3). In the winter months, average wind speeds increase slightly (avg. ca. 5-6 knots), with gusts averaging 11-14 knots and periodic wind speeds reaching much higher velocities. Between November and February, the trade winds blow from the east-southeast. During winter months, winds are more variable. As weather fronts pass through, the wind direction spirals around the island over a period of 4-8 days, resulting in the more-evenly distributed wind pattern seen in Figure 3.

No reliable wave data were available, but it can be reasonably assumed that the approach patterns of local seas are similar to the winds summarized in Figure 3. During August and September of our two field seasons, wave heights above 5 meters were common along the more exposed shore, which varied with wind direction. The most influential waves, however, are long-period Antarctic swell from the south and southeast. On one occasion, swell with a wavelength of 1 - 1.5 km was observed off Rano Kao and presumably would have "felt bottom" to a depth of ca. 750 m. As will be discussed below, this is one of the primary controls of coral distribution around the island.

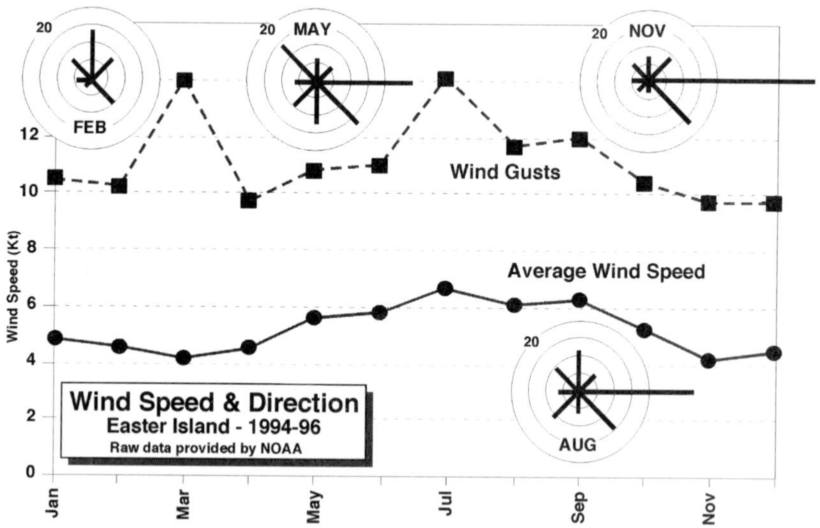

Figure 3. Wind speed and direction on Easter Island. The speeds are monthly averages taken between January 1994 and December 1996. The length of each bar in the circular wind roses reflects the percent of time that the wind blew from that direction. Four examples are provided to illustrate seasonal wind patterns. Summer winds (e.g., November) are strongly affected by the easterly trade-wind circulation. In the spring months (e.g., August), day-to-day winds average 1-2 knots faster than during the summer, and much higher winds can blow on a particular day. In the winter months (e.g., February), stronger winds are associated with passing fronts, and wind direction is highly varied, typically spiraling around 360° over a period of 4-8 days.

3. PREVIOUS CORAL-REEF STUDIES

Because of the cold water temperature, the harshness of the wave climate and the distance of Easter Island from known sources of coral larvae, coral diversity has been depressed and coral abundance has been assumed to be likewise low. In 1972, Wells reported 6 scleractinean species from Easter Island. DiSalvo, *et al.* (1988) increased this to 11 species.

A quantitative survey in the mid-80's reported 0.04% cover by *Pocillopora damicornis* and 0.34% cover by *Porites lobata* near Anakena (DiSalvo, *et al.,* 1988). Based on nearly 40 scuba dives in 1985 and 1986, they noted a modest increase in coral cover below a depth of 10 m, but concluded that the low cover measured along the Anakena transect "appeared representative of the average coral growth on the island at this time" (p.456).

4. METHODS

In August of 1998, 13 reconnaissance dives were made using scuba along all three shelves to a maximum depth of 45 m (Fig. 1; squares). The intent was to provide general information on the presumably depauperate coral community and its relationship to the varying physical-oceanographic regime around the island. The primary record of the 1998 excursions consisted of observations recorded on underwater slates and panoramic video shot with a Sony Hi-8 camera in an underwater housing. Despite a lack of quantitative measurements, the impressions of the reef were dramatically different from those reported previously. While diversity was low (two species dominated), most dives revealed abundant coral fauna, in some instances approaching 100% cover over large areas (ca. 200-500 square meters).

Based on these preliminary observations, a follow-up diving program was initiated in 1999. Its objectives were to: 1) better document the patterns of coral occurrence around the island, 2) provide at least semi-quantitative measurements of coral abundance, and 3) relate those patterns to prevailing oceanographic conditions. In August of 1999, 18 dives were made to a maximum depth of 52 m (most dives reached maximum depths of 30-35 m), again circumnavigating the island to observe the benthic communities on all three coasts (Fig. 1). Detailed field notes were taken to record the character of the nearshore profile and patterns of coral cover, fleshy algae, grazing urchins and other benthic macrofauna. Underwater video was shot with a Sony TRV-900 digital camera in a Gates housing, taking care to regularly record water depth on the tape. At roughly 3 m depth increments, 20 m video profiles were run along the approximate depth contour, keeping the camera vertical and 80-100 cm off the bottom. Still photographs were taken of important features using a Nikonos V underwater camera with a distortion-corrected 20-mm lens. The water depth of each photograph was recorded for comparison to the video and field notes. The descriptions below are based on qualitative evaluation of the video transects, still photographs and field notes. Detailed analyses of coral abundance are underway and will be presented in a later paper.

5. RESULTS

5.1 Coral Cover

While coral cover is variable, it is generally much higher than previously reported. Diversity is low, and the coral community is dominated by two species, *Porites lobata* and *Pocillopora verrucosa*. DiSalvo, *et al.* (1988) identified most shallow species as *P. damicornis*; Peter Glynn (University of Miami, personal communication) identified the most common shallow species as *P. verrucossa*, and we defer to the latter identification. *P. damicornis* and *P. eydouxi* were also seen but in lower abundance (fewer

Figure 4. Underwater photograph of abundant colonies of *Pocillopora verrucosa* at a depth of 6 meters (19 ft) off the low-energy northern coast near Hanga Otea (Fig. 1). At other locations where wave energy is lower, coral cover starts at shallower depths. Photo by Henry Tonnemacher.

Figure 5. Underwater photograph of a large colony of *Porites lobata* off Anakena (Fig. 1). Note the skirts forming near the base of the colony at a depth of 17 meters (55 ft). Photo by Henry Tonnemacher.

than one in 1,000 colonies for *P. eydouxi*). *Pocillopora* spp. generally dominate in shallower water (<10-15 m) on horizontal substrates (Fig.4). *Porites lobata* typically starts at a depth of 10-15 meters and is more abundant along vertical and sloping surfaces. In shallower water its morphology is dominantly hemispherical. Dome-shaped colonies exceeding 2 meters in diameter were seen at sites along all three coasts. In particular, one large colony nearly 10 meters across was encountered off Anakena along the north shore (Fig. 5). Starting at 15-20 m, colonies become more conical (Fig. 6). It is common for digitate features to project upward from inclined colony surfaces at intermediate depths. DiSalvo, *et al.* (1988) reported the conical morphology but characterized it as being limited to deeper, protected areas. In contrast, we found this colony morphology to be ubiquitous in all areas not directly impacted by southeasterly swell. In deeper water (>20 m), colonies develop plate-like skirts, much like those seen in Caribbean *Montastraea* sp. that were experimentally transplanted from shallow to deep water by Graus and Macintyre (1976).

5.2 Coral Distribution

The type and abundance of corals around the island is strongly dependant on water depth, bottom slope and exposure to wind waves and oceanic swell.

At most sites, coral cover is largely absent in water depths less than 5-7 meters (Fig. 7). Immediately below that depth, *Pocillopora* spp. generally dominates the bottom (Fig. 4). *Pocillopora* cover is consistently higher than that reported by DiSalvo, *et al.* (1988), except along the southeast-facing coast. Below a depth of 10 meters, *Porites lobata* progressively dominates with increasing water depth. In most areas along the northern and western coasts, cover by *Porites lobata* is high, and combined cover by *Porites* and *Pocillopora* spp. averaged near 50%. Total cover approached 100% over areas larger than 300 m^2 at depths between 20 and 50 meters.

Figure 6. Conical heads of *Porites lobata* off Anakena (Fig. 1) at a depth of 30 meters (100 ft). Photo by Henry Tonnemacher.

5.3 The Occurrence of Reefs

It has been generally held that "coral reefs do not exist at Easter Island". If we use definitions that require the structure to reach sea level, then this remains the case. If the criteria center around topographic structures creating relief by multiple generations of coral growth (i.e., not simple veneers on previously existing topography), then examples of reefs do exist.

Figure 7. Barren volcanic rock at a depth of 3 meters (10 ft) off Cabo O'Higgins (Fig. 1).

In deeper water, much of the topography represents a veneer of live coral on top of lava flows and tubes oriented perpendicular to shore (Fig. 8). While superficially similar to spur-and-groove structures in the Caribbean, these features simply mimic the underlying volcanic terrain. However, along the north side of the Poike Peninsula near Hanga Tavaka (Fig.1), much of the topography appears to have developed independently from the underlying substrate. The carbonate cap is not a simple veneer of living corals, but rather a build-up composed of multiple generations of both in-place and dislodged corals, held together by encrusting algae, submarine cement and overgrowth by new corals (Fig. 9). Likewise, off the western shore the bottom topography seems to be dictated in places more by coral development than the shape of the underlying substrate. At Punta Cook, on the west coast, a large colony of toppled *Porites lobata* was found (Fig. 10a). This exposed a clearly coral-dominated reef interior (Fig. 10b). While these features do not reach sea level, they are similar to coral-built structures found along many Caribbean and Pacific shelf breaks.

Figure 8. High coral cover on a lava tube off Motu Tautara on the west coast (Fig. 1). The morphology is similar to that seen in spur-and-groove topography in Caribbean reefs. However, the latter are entirely constructed by coral growth, whereas this example is simply a veneer of coral over preexisting topography related to a previous lava flow. Photo by Henry Tonnemacher.

Figure 9. Underwater photograph of a possible coral reef at a depth of 23 meters (74 ft) off the north side of the Poike Peninsula (Fig. 1). The topographic structures at the left are formed by multiple generations of *P. lobata* and are not simple veneers over antecedent topography. Photo by Henry Tonnemacher.

Figure 10. TOP. Underwater photo of a toppled colony of *P. lobata* at a depth of 21 meters (70 ft) off Punta Cook (Fig. 1). *BOTTOM.* Photograph of the reef interior exposed behind the toppled *P. lobata* colony shown in A. Note the in-place skirts of previous colonies exposed in the cut. This building by multiple generations of coral is central to separating reef buildups from coral veneers.

5.4 Controls of Reef Development

Figure 11 summarizes information on near-shore bathymetry, coral abundance, and the occurrence of non-calcareous algae and grazing urchins around Easter Island. In general, the insular shelf narrows and steepens near the volcanic cones at the three corners of the island. Because sediment runoff from land tends to be diverted to the lower coastal areas in between, conditions for coral growth are generally more favorable off steeper, cliffed coastlines.

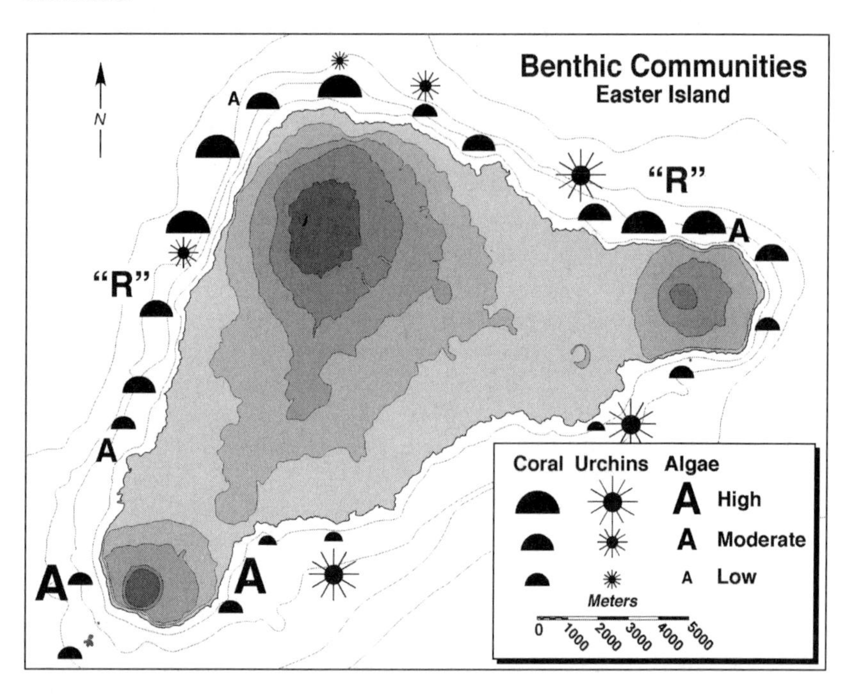

Figure 11. Summary of coral, urchin and algal abundance at dive sites around Easter Island. The main control of coral cover is wave action. Waves related to the easterly trade winds and refracted/diffracted (e.g., reduced energy) swell create conditions more favorable to coral and reef development. As a result, the most abundant coral cover occurs along the northern and western insular (i.e., leeward) shelf. Reefs are discouraged on the southeastern coast by long Antarctic ground swell that prevents corals from settling or reaching any significant size. Urchins (e.g., *Diadema*) are highly variable in their abundance and are generally either absent or very abundant (i.e., few instances of intermediate density). Where urchin densities are high, the bottom is largely devoid of algae, and often coral. The areas identified as possible "reefs" (R) reflect build-ups of multiple coral generations under a sustained balance between scleractinean coral, algae and grazing urchins.

Areas of highest coral abundance are characterized by moderate to low levels of fleshy algae and grazing urchins, in particular species of *Diadema*. *Halimeda* spp. (calcareous algae) and *Lobophora* spp. (fleshy algae) were observed in lesser quantities where coral cover was high, and were usually confined to areas between corals or beneath overhangs. Along the southeastern coast, taller populations of a *Sargassum*-like alga tend to preclude abundant corals even in protected areas (e.g., Hanga Vinapu, Fig. 1). One of the authors (MG) has noted over the past 20 years that the urchin populations tend to move from site to site around the island, decimating the algal population at one locale and then moving on. Highest coral cover was usually associated with a modest community of short, fleshy algae and a lesser numbers of grazers. In this vein, it is worth mentioning that parrotfish and other scarids have been virtually absent from Easter Island since a massive die-off of algae after the 1982-83 El Niño.

The primary control on coral abundance is exposure to severe wave action and, in particular, Antarctic ground swell. While wind waves regularly create conditions that are too rough for safe navigation, long-period swell seems to provide the dominant limitation for coral development. The north coast is dominated by trade-winds circulation throughout the summer months. While the areas along the north and west coasts receive heavy surf during the passage of major weather fronts, wavelengths associated with local seas are generally shorter and the impact on deeper bottom communities is, therefore, much less than that associated with longer-period swell. In 1998, southeasterly swell was observed with a wavelength of nearly 1.5 km. These waves would have "felt bottom" at depths up to 750 meters and caused strong surge all along the southeasterly shelf. Such conditions are probably not rare.

The southeast coast between Orongo and the southern Poike Peninsula faces directly into the southerly and southeasterly swell, and coral cover is lower. Along the exposed, vertical sides of Motu Nui and Mutu Iti (at the southern tip of the island), coral cover is absent above a depth of 15-20 meters, even in protected areas where wave energy is still high (Fig. 12).

Seven dives along the southeast-facing coast in 1998 and 1999 revealed a wave-swept bottom dominated by barren basalt flows (Fig. 13) except for areas where bottom topography provided localized shelter from wave attack (e.g., landward-facing vertical surfaces). At more exposed sites, the local urchin population is confined to those species that are capable of burrowing into the lava (e.g., *Echinometra, Echinostrephus*). Where protection from wave action is afforded, the long-spined sea urchin (*Diadema* sp.) can be common.

Figure 12. Underwater photograph off Motu Iti (Fig. 1) taken from a depth of 18 meters (60 ft). Note the generally poor coral cover even at greater depths in areas exposed to direct attack by southeasterly swell. Photo by Henry Tonnemacher.

6. DISCUSSION

6.1 General Trends

The coral population around Easter Island understandably lacks diversity due to cold water, heavy surf and the considerable distance from possible larval pools. The abundance of corals is surprising, however, and is in dramatic contrast to previous reports from Easter Island. It appears that in the absence of competition by other species, *Porites lobata* and *Pocillopora* spp. have occupied a significantly larger area than was previously thought to be the case. While successful recruitment may occur on hard substrates over a wide variety of physical-oceanographic conditions, survival to adulthood is tied closely to wave energy, and in particular to the strength of oceanic swell.

Figure 13. Barren lava surface at a depth of 10 meters (33 ft) off Hanga Vinapu. Scour by southeasterly swell make it impossible for coral recruitment. The modest urchin population has removed all algae at this exposed site.

This larger-scale pattern of coral development is directly opposed to most generally accepted ideas. James (1983) succinctly summarizes the well-established model of coral-reef development: "they (reefs) are best developed and most successful on the windward sides of shelves, islands and platforms where wind and swell are consistent and onshore." On Easter Island, the best-developed (and perhaps only) areas of significant coral cover occur on leeward shores, away from the dominant approach of Antarctic swell.

To the north, well-developed coral assemblages thrive under the influence of the easterly trade winds during the summer months. While storm waves might be high in this area, their shorter wavelength makes them less effective than southeasterly swell. Along the western shore, the dominant wave action comes from swell refracted and diffracted around the southern tip of the island. While wave energy is lower than along the southeast coast, swell nevertheless exerts an important although indirect effect on shelf processes. The occasional strength of wave action even along leeward shores is reflected in the toppled colonies along the west coast (e.g., Fig. 10), as well as the large, rounded boulders and large ripples seen off Motu Tautara (Fig. 14).

Figure 14. TOP. Large ripples on a sandy bottom at a depth of 18 meters (60 ft) off Motu Tautara on the west coast (Fig. 1; Photo by Henry Tonnemacher.). *BOTTOM.* Large, boulders off Mutu Tautara reflecting very large waves that occur often enough to round their surface. The western shelf is partially protected from heavy swell, but wave energy remains high enough to move sediment on a regular basis.

The relegation of reefs to sites that are more protected from intense wave action is not a new observation. Grigg (1997) described a similar situation in Hawaiian reefs, where Pacific swell prevents reef development to windward. Likewise, many raised Pleistocene reefs in the Caribbean Sea have been reported from the western sides of the Windward Islands. What remains most surprising on Easter Island is the high abundance of corals in what has traditionally been considered to be a severely compromised environment.

6.2 Differences from Previous Reports

An obvious question concerns the discrepancies between this and earlier studies. DiSalvo, *et al.* (1988) reported island-wide coral cover on the order of a few percent, although they did allow for higher but patchy cover at greater water depths. They pointed out a decline in algae such as *Sargassum* spp. compared to levels observed in the late 1960's. Whether these losses are related to a population explosion by *Diadema* as they proposed or to environmental fluctuations related to El Niño in the mid 1980's cannot be determined from available data. Nevertheless, a dramatic shift in shallow-water communities undoubtedly occurred around this time. Video shot by Jacques Cousteau shows extensive *Sargassum* meadows in shallow water prior to 1980. By contrast, this alga was observed only in a few places in 1998 and 1999 (e.g., Hanga Vinapu), and even there in lower abundance and in stunted condition. The near disappearance of this once-dominant alga undoubtedly opened up new space for coral recruitment in shallow water where *Pocillopora* dominated in 1998 and 1999.

One of the authors (MG) has been involved in a long-term monitoring program for the fuel pipeline at Hanga Vinapu. Based on comparisons of coral cover in older video to what we observed, it seems reasonable that levels of *Pocillopora* in August 1999 could have recruited since the study by DiSalvo. Video surveys of several hatch covers from a recent shipwreck off Hanga Roa show a similar pattern (modest recruitment by *Pocillopora* where algae have not covered the substrate).

After algal losses in 1982-83, *Pocillopora* recruitment in shallow water would not have advanced sufficiently by 1985 and 1986 to produce the coral abundances observed in 1998-99. It therefore, seems realistic that the discrepancy between the shallow-water observations of DiSalvo, *et al.* (1988) and this paper can be explained by the recruitment of primarily *Pocillopora verrucosa* since the earlier surveys were completed. As an aside, locals lament the loss of algae that have been displaced by recent coral recruitment. This is in sharp contrast to world-wide concerns over coral loss and algal overgrowth in most tropical regions. One's perception of a "healthy" reef often reflects what was there when it was first visited.

Discrepancies between high coral cover found in deeper-water in 1998-99 and the very low cover reported by DiSalvo, *et al.* (1988) cannot be similarly explained. While they did allow for higher cover in deeper water, the coral abundance reported for 1985-86 was clearly far below anything we observed in 1998-99. Given the large size of the colonies observed at most sites and the known growth rates of *Porites* (ca. 1 cm/yr: Isdale, 1977), the high population densities observed in the 1998 and 1999 surveys would have also been present at the time of the earlier survey.

The high abundance of corals at all depths raises questions about how the population is maintained, given the dominant water flow to the west and away from Easter Island. Based on remote-sensing data, Lagerloef *et al.* (1999) proposed that eastward-flowing currents dominate the equatorial Pacific during major El Niño events. If this is the case, then coral larvae from the Marquesas and Tuomotu Islands or even the core Indo-Pacific region could reach Easter Island during years in which ENSO events have occurred. If genetic information could show an origin for local corals in the Marquesas Islands, then it would show that larvae and Hotu Matua, the island's first inhabitant, perhaps share fortuitous circumstances and a common drift pathway to Easter Island.

6.3 Human Impact

The small size of the island and the narrow shelf make development-related impact inevitable. Runoff from the island interior is undoubtedly exacerbated by agriculture on slopes that were denuded of nearly all larger vegetation many years ago. Water clarity off Hanga Roa is noticeably below that of adjacent areas away from town and port activities. Likewise, runoff from the unpaved road to Hanga Vinapu stresses an environment already taxed by heavy oceanic swell.

Impact from the vigorous harvesting of *Pocillopora verrucosa* for sale to tourists appears to have been surprisingly low. On two occasions over a 4-week period, small boats brimming with hundreds of harvested colonies were observed landing at Hanga Roa and Hanga Piko. Apparently, the colonies were usually taken from areas of high cover and this practice, combined with rapid recruitment resulted in lower-than-expected impacts. The recent near-extirpation of *Pocillopora* in shallow water (discussed below) will make it more difficult to harvest the genus, even in small numbers, without significant impact. Colonies of *Pollopora eydouxi* were also seen on sale in the public market. Because this coral is rare, its harvest undoubtedly has deleterious, although unquantified effects.

The greatest impact that was observed during the1999 survey occurred in deeper coral gardens off Anakena, where large supply vessels anchor to wait out periods of heavy southeasterly swell. Large areas of recently devastated colonies were observed on a single dive made on 25 August, 1999 (Fig. 15). The level of damage is clearly greater than anything related to coral harvesting and involves larger colonies that probably took close to a century to form. Some means of reducing anchor damage should be considered. While permanent moorings would require frequent maintenance, they could preclude coral damage altogether. At a minimum, a designated anchorage further offshore in an area of low cover or open sand should be instituted. As areas of lower coral cover are not far away, it is unlikely that this would significantly compromise the safety or comfort of the vessels seeking shelter from heavy seas.

Figure 15. Toppled colonies of *P. lobata* at a depth of 25 meters (80 ft) off Anakena (Fig. 1). Damage is caused exclusively by large vessels anchoring in the area to escape heavy weather from the south.

6.4 What is a "Healthy" Reef?

While the nearshore marine community seemed robust based on the number of corals present, low diversity leaves the system at risk. If exposed to natural or anthropogenic stresses that are adverse to either *Porites lobata* or *Pocillopora verrucosa*, the reef system lacks the diversity necessary to offset any impact. This is well illustrated by the sudden and widespread bleaching that occurred on Easter Island in March and April of 2000. Because Easter Island corals live in colder water, they are not normally subjected to higher water temperatures associated with extensive bleaching and coral mortality in warmer tropical areas. Logically, this would make them less susceptible to the epidemic levels of bleaching noted elsewhere around the world.

However, in February 2000, water temperature reached higher-than-normal levels and remained elevated through late March (dashed line in Figure 2). In early March, corals in shallow water bleached extensively (Fig. 16a) and remained so for several months. The greatest impact involved the *Pocillopora* community, which was reduced by 90 percent or more in water depths less than 10-15 meters. At greater depths, *Porites lobata* showed variable bleaching (Fig. 16b) and little or no damage was seen below 25 meters. Similar bleaching has occurred in the past, but was associated with cooler winter temperatures and attributed to elevated rainfall (Cea Egaña and DiSalvo, 1982).

The expulsion of photosynthetic zooxanthellae apparently occurred at temperatures well below those associated with bleaching elsewhere. Also, warming was associated with a La Niña cycle, in contrast to northern tropical areas where El Niño-associated warming is more the norm (Wellington, *et al.*, 2000). The Easter Island example infers that corals acclimate to their ambient temperature and bleach at a level that is some percentage above normal rather than an absolute threshold. This is bad news for those who hoped that corals in cooler water might represent a "population reserve" from which larvae could repopulate into areas damaged by thermal stress. In contrast, it places the coral population around Easter Island at perhaps even greater risk than other reefs in warmer water where coral diversity is higher.

The scenario at Easter Island raises some important issues about what constitutes a "healthy" reef system. If observed out of context, the abundant corals observed in 1998 and 1999 would prompt most researchers to conclude that this reef was robust and well suited to local conditions. However, the reef community, and in particular that in shallower water, has undergone some significant and rapid changes in only two decades. Prior to the 1980's, the shallow shelf was dominated by luxuriant macroalgal meadows that supported a community of grazers not found on Easter Island today. After 1982 and 1983 (years associated with an El Niño event), the algae virtually disappeared and were replaced in less than a decade by a coral community dominated by *P. verrucosa*. Recruitment created a shallow-water coral community that was sparse in 1985 but increased significantly over the next three years. Then, in 2000 the La Niña-associated warming induced widespread bleaching that decimated the *P. verrucosa* community that had achieved abundances approaching 100 percent over large areas only 2-4 years earlier.

Figure 16. Recent coral bleaching near Motu Kau Kau on the southern tip of Easter Island (Fig. 1). *TOP*. Bleached colonies of *Pocillopora verrucosa* at a depth of 8 meters (26 ft). *BOTTOM*. Bleached *Porites lobata* at a depth of 15 meters (48 ft).

Which of these communities was “normal”? What constitutes a “healthy” reef community in this instance? Is one of the community structures observed between the early 1970’s and today the one that is “supposed to be there”... and the rest aberrations? Or are the rapid shifts that have been the hallmark of the past two decades simply part of a larger boom-and-bust cycle in which wild changes in community structure are to be expected? And, regardless of which (if any) is the case, to what extent might local stresses related to infrastructure development on Isla de Pascua and larger-scale changes in global climate stability be playing a role? The reefs of Easter Island remain as much of an enigma in the marine system as do the toppled moai on land. The unknown cultural rise and fall on the island probably represents a microcosm of the interplay between increased anthropogenic pressure and environmental degradation on a global scale. Likewise, answers to questions about why reefs form where they do and what factors they respond to on a variety of spatial and temporal scales may reside in the surprisingly rich but variable marine community that exists offshore.

7. CONCLUSIONS

Based on our observations and preliminary measurements, the following conclusions are offered:

1. Cold water, a generally harsh wave climate and great distance from known centers of coral diversity have resulted in a coral population that is represented by a limited number of species.
2. However, coral abundance is surprisingly high, averaging over 50% along the northern and western coasts at water depths between 10 and 35 meters.
3. In shallow water, several species of *Pocillopora* (primarily *P. verrucosa*) dominated where wave energy is low enough to allow successful larval recruitment and continued coral growth. These opportunistic species tend to favor horizontal substrates.
4. At depths below ca. 10-15 meters, *Porites lobata* increasingly dominates. At the shallower end of this range, large hemispherical colonies can reach diameters exceeding 8 meters (e.g., Fig. 5). At greater depths (20-30 m), colony shape becomes more conical (e.g., Figs. 6 and 8) and eventually more plate-like. *Porites* occurring in shallow water is more common on sloping substrates.
5. In many areas deeper than 20 meters, coral cover averages above 50% with large patches characterized by nearly 100% cover, predominantly

Porites lobata. The greatest coral abundance occurs in areas with low cover by *Lobophora* and moderate urchin (e.g., *Diadema* sp.) densities.

6. Coral cover is best developed along leeward shelves away from the direct impact of long-period, southeasterly swell. The southeast-facing shelf is typically a barren and eroded lava pavement that supports a variable population of burrowing urchins. Coral is found mostly as refugia and is generally dominated by *Pocillopora* spp. and *Porites lobata*. The apparent preference for leeward sites is in direct contradiction to popular models that stress reef development being confined to windward margins.
7. What might be described as true reefs were found off the north side of the Poike Peninsula and near Punta Cook (Figs. 6 and 10) on the west coast. Both areas are impacted by refracted swell, but do not receive the full force of these long-period waves. Likewise, both areas are characterized by moderate populations of fleshy algae and grazing urchins, perhaps reflecting a long-term balance of corals, algae and grazers.
8. Differences in the abundance of shallow-water corals reported by DiSalvo, *et al.* and this study can be explained by recruitment of *Pocillopora* after the extirpation of *Sargassum* in the early to mid 1980's. Recruitment rates on substrates of known age are consistent with this hypothesis.
9. Earlier reports of low abundances within the deeper reef community cannot be reconciled by such a scenario. The large *Porites lobata* colonies seen in 1988-89 were undoubtedly present in 1986 and it is unclear why they were not observed.
10. While human influence is less dominant than might be expected on an island with such a narrow shelf, it still remains an important factor. Because the nearshore coral community lacks the diversity to mitigate major losses of *Porite lobata*, *Pocillopora verrucosa*, or any other dominant species, elevated natural or anthropogenic stress could pose significant threats to the entire nearshore marine community.
11. The fragility of the marine system around Easter Island is highlighted by the recent coral-bleaching event in early 2000. Corals expelled their zooxanthellae in response to elevated temperatures that are below levels associated with bleaching elsewhere. This may reflect a bleaching response that is more related to increased temperatures that are some percent above ambient levels, rather than an absolute upper thermal threshold.
12. The highly variable community structure of the shallow shelf around Easter Island raises interesting and important questions about what constitutes a "normal" or "healthy" marine community. Furthermore, Easter Island may be an ideal place to further examine these issues.

ACKNOWLEDGMENTS

To a large extent, this study has been an exercise in which one author (DKH) has systematically documented what the other (MG) has known for years. The effort was made possible through the support of the Science Museum of Long Island, the U.S. National Park Service and the Explorers' Club. In particular, we wish to acknowledge the considerable efforts of Dr. John Loret, who used his earlier association with Thor Heyerdahl as a catalyst to make the more recent expeditions a reality. Before his move to Dowling College, Dr. John T. Tanacredi coordinated the National Park Service's participation in 1998 and organized the entire expedition in 1999. His research into the limits of equine loading will no doubt loom legendary. We also gratefully acknowledge the considerable efforts of Sergio Rapu, a lifetime resident of Easter Island and its senior archaeologist. His insight into local customs and island history were invaluable, as were his friendship and personal support for the expedition and this project.

Our field efforts were greatly enhanced by several individuals without whose assistance the work would have been impossible. Our small "research vessel" was ably captained by Eugenio Hey Riroroko, who safely guided us into and out of what many might not consider to actually be harbors. The 1998 survey benefited greatly from the observations and assistance of (in alphabetical order) Norm Baker, Gerry Bunting, Tobey Curtis, Emily Loose and Rich Pekelney. Their companionship and thoughtful comments added significantly to the impressions gained during the first survey. And finally, Henry Tonnemacher is very gratefully acknowledged for his considerable assistance during the 1999 survey, including the careful photo-documentation that resulted in many of the underwater photographs appearing in this chapter. His association with the first author has spanned over two decades, and what has become an almost "autopilot" underwater relationship made for a smooth field effort.

REFERENCES

Cea Egaña, Alfredo, and DiSalvo, L.H., 1982, Mass expulsion of zooxanthellae by Easter Island corals, *Pacific Science*, 36: 61-63.

DiSalvo, L.H., Randall, J.E., and Cea, A., 1988, *National Geographic Research,* 4: 451-473.

Druse, S., Duncan, R.A., Liu, Z.J., and Naar, D.F., 1997, Effective Elastic Thickness of the Lithosphere Along the Easter Island Seamount Chain, J. Geophysical Research, B, Solid Earth and Planets, 102 (no. 12): 27, 305-27, 317.

Graus, R.R., and Macintyre, I.G., 1976, Light Control of Growth Form in Colonial Reef Corals: Computer Simulation, *Science,* 193: 895-897.

Grigg, R.W., 1997, Paleoceanography of Coral Reefs in the Hawaiian-Emperor Chain Revisited, *Coral Reefs*. 16: 33-38.

Isdale, P., 1977, Variation in Growth Rate of Hermatypic Corals in a Uniform Environment, Proc. Third Intl. Coral Reef Symp., 2: 403-408.

James, N.P., 1983, Reef Environment in: *Carbonate Depositional Environments* (Scholle, P.A., Bebout, D.G., and Moore, C.H., ed.), 346-462.

Lagerloef, G.S., Mitchum, G.T., Lukas, R.B., and Niiler, P. p., 1999, Tropical Pacific Near-surface Currents Estimated from Altimeter, Wind and Drifter Data, J. Geophys. Res., 104: 23, 313-23, 326.

Newman, W.A., and Foster, B.A., 1983, The Rapanuian Faunal District (Easter and Sala y Gomez): In search of Ancient Archipelagos. Bull. Mar. Sci., 33: 633-644.

Randall, J.E., and McCosker, J.E., The Eels of Easter Island with a Description of a New Moray, *Contributions in Science*, No. 264, Natural History Museum of Los Angeles County, Los Angeles, CA.

Wellington, G.M., Glynn, P.W., Strong, A.E., Navarrete, S.A., Wieters, E., and Hubbard, D.K., 2000, Crisis on coral reefs linked to climate change, *EOS*, 82: 1-5.

Wells J.W., Notes on Indo-Pacific Scleractinean Corals, pt. 8: Scleractinean Corals from Easter Island, *Pacific Science* 26: 184-190.

PART II

ECOLOGICAL CONSIDERATIONS AND RESTORATION/PROTECTION EFFORTS OF NATURAL AND CULTURAL RESOURES

SMLI/NPS supported SCUBA divers on a typical coral head along the coastline of Easter Island in approximately 220ft of water. Coral cores were taken from these coral formations.

Chapter 5

Rapid Vegetational and Sediment Change from Rano Aroi Crater, Easter Island

DOROTHY PETEET[1,2], W. BECK[3], J. ORTIZ[4], S. O'CONNELL[5], D. KURDYLA[2], AND D. MANN[6]
[1]NASA/GISS, 2880 Broadway, NY, NY 10025; [2]Lamont Doherty Earth Observatory, Palisades, NY 10964; [3]AMS Facility, Dept. of Physics, University of Arizona, Tucson AZ 85721; [4]Kent State University, Kent State, OH; [5]Dept. of Earth and Environmental Sciences, Wesleyan University, Middletown, CT; [6]Institute of Arctic Biology, University of Alaska, Fairbanks, AK 99775

1. INTRODUCTION

Previous pollen investigations by Flenley *et al.* (1991) suggest that the Rano Aroi Crater, Easter Island (27 08' S, 109 26'W) contained a record of vegetational history that is greater than 35,000 C-14 years old. The environmental setting and modern vegetation is fully discussed by Flenley *et al.* (1991). Rano Aroi Crater was selected for further work for two reasons. The first is that we were interested in a sediment record that spanned the transition from glacial to interglacial conditions. The second is that the peat composition might prove ideal for macrofossil analysis. Large fluctuations in pollen percentages characterize the stratigraphy, suggesting that this small subtropical island showed a major vegetational response to climate change. In order to add insight into the vegetational and sediment changes that took place, we re-cored the Rano Aroi Crater and are in the process of analyzing the core at 2-cm intervals for pollen and macrofossils in addition to sediment characterization and AMS C-14 dating. Our preliminary results here focus on the most striking lithological change in the core, Drive 3, between 200 and 300 cm depth.

Easter Island, Edited by John Loret and John T. Tanacredi
Kluwer Academic/Plenum Publishers, New York, 2003

2. METHODS

2.1 Core Acquisition and Pollen/Macrofossil Analysis

In August, 1997, an 8-m core was collected about 49 m from the edge of the Rano Aroi Crater (elev. 425 m) using a 5-cm diameter modified Livingstone piston corer (Wright *et al.*, 1984). Some of the drives were shorter than the 1 m recovery due to sediment compaction. The cores were extruded, wrapped in saran wrap and foil, and then shipped to the US for analysis. We sampled at the 2-cm resolution for pollen and spores, but eight initial samples throughout the interval of interest are presented. Pollen processing followed Faigri and Iversen (1975) with modifications by Heusser and Stock (1984) including sieving with 150 micron and 7 micron screens. A known quantity of Lycopodium tablets was added to calculate pollen accumulation rates and pollen influx. Pollen and spore residues were mounted in silicone oil to facilitate rotation. Pollen identification followed Flenley *et al.* (1991). A sum of at least 100-200 upland pollen grains is presented in the diagram for each sample, but final results will total 300. Pollen percentages are presented using total pollen for a sum; spore percentages use total pollen and spores as a sum.

Exotic Lycopodium percentages are not included in the sum, but calculated as a percentage of all pollen and spores and are presented to illustrate the differences in amounts of pollen accumulation. Macrofossils were screened from the bulk sediment using 500 and 250 micron screens. Macrofossils are not presented due to the paucity of identifiable remains in the samples screened thus far.

2.2 AMS C-14 Dating

Radiocarbon measurements were made at the University of Arizona AMS Facility. The materials dated were plant macrofossils, which were mechanically separated from the core. These materials were pretreated with a standard HC1/NaOH/HC1 pretreatment for 72hr/1hr/24hrs respectively at 70° C to remove soil carbonate and humate material which may have accumulated while in the sediment column. Samples were then combusted at 400°C in 0_2 gas using a temperature controlled furnace (Jull *et al.*, 1998). A correction for stable isotope fractionation was made based on the measured delta C^{13} (Donahue *et al.*, 1990). Ages are standard radiocarbon ages using the Libby half life. Quoted errors are 1 standard deviation, and include terms for counting statistics, machine error, and uncertainty in the processing blank.

2.3 Diffuse Spectral Reflectance (DSR)

Sediment color for Rano Aroi Drive 3 (200-300 cm depth) of Easter Island was quantified using a Minolta CM-2022 spectrophotometer (spotsize – 4mm). This feasibility study was conducted to assess the potential of Diffuse Spectral Reflectance to provide both high-resolution stratigraphy and insights into sediment composition in terrestrial (crater) environments. DSR measurements were obtained along the 84 cm long section at 1 cm resolution after lightly scraping the core to smooth the surface and remove any potentially oxidized material from the surface layer. The core surface was then wrapped in Gladwrap to protect the instrument's integration sphere and prevent moisture loss from the sediment.

3. RESULTS

3.1 Lithology and AMS C-14 Dating

The lithology of the Rano Aroi, Drive 3 section, representing the interval from 200 – 300 cm depth in the crater (84 cm in length retrieved after compression) is as follows:

Depth	Lithology Description
0-39 cm	Brown fibrous organic peat
40-55 cm	Greyish-brown organic clay
56-66 cm	Greyish-brown organic clay with fibrous material
66-67 cm	Grey clay
67-84 cm	Brown organic peat

Initially, 3 samples of the core were screened for macrofossils and sent to the University of Arizona. These results suggested that Drive 3 contained the late-glacial interval, beginning around 13,605 ± 90 years ago and extending into the early Holocene (9,650 ± 170) (Fig. 1). The striking changes in lithology within this 30-84 cm section of the core were our initial reason for focusing in more detail on the pollen and sediment stratigraphy. We initially rejected the age of 1,944 ± 42 in Drive 3, 60-62 cm because of possible contamination.

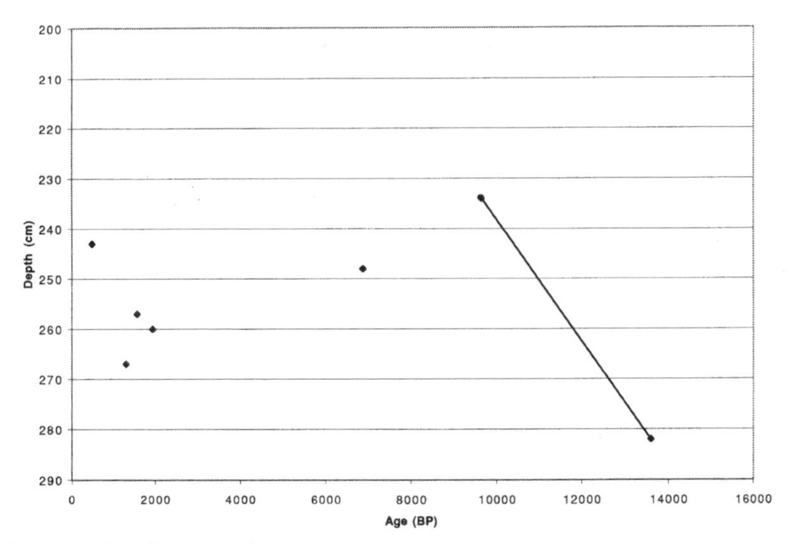

Figure 1. AMS C-14 ages plotted against depth in the Rano Aroi Drive 3 core, 200-300 cm depth. We hypothesize that the correct age is the late-glacial Holocene transition interval (line on graph), which corresponds with Flenley *et al.* 1991 stratigraphy in the crater.

1998 Samples

AA LAB NO	SAMPLE ID	MATERIAL DATED	C13 VALUE	C14 AGE
AA29091	RA2-drive3	34-36 cm Spongy bark	-25.57	9,650±170BP
AA29090	RA2-drive3	60-62 cm Stringy red epidermis	–25.77	1,944±42BP
AA29089	RA2-drive3	82-84 cm Stringy red epidermis	–27.25	13,605±90BP

Subsequent macrofossil analysis of the following 2-cm samples yielded primarily reddish epidermal material, which we sent for further AMS dating. The 2001 results are all younger than the previous 1998 dates, and suggest that the reddish epidermal material is partially root contamination of the sediments. Alternatively, it is possible that the 1998 late-glacial ages are contaminated from older carbon washing into the crater.

2001 Samples

AA LAB NUM	SAMPLE ID		C13VALUE	FMODERN	C14 AGE
AA43217	Z1928A	RA-2, drive-3, 42-43 cm	-27.1	0.9406±0.0051	492±43
AA43218	Z1929A	RA-2 drive-3, 48-49 cm	-25.0	0.425±0.012	6,880±230
AA43219	Z1930A	RA-2, drive-3, 57-58 cm	-27.1	0.8230±0.0034	1,565±33
AA43220	Z1931A	RA-2, drive-3, 67-68 cm	-25.1	0.7850±0.0041	1,300±100

3.2 Diffuse Spectral Reflectance

Mean DSR measurements (400-700 nm) are presented in Fig. 2. Triplicate measurements at the base of the section indicate errors that range from $\pm$0.5 to 1.5% over the spectral range of the instrument (400-700nm), which yields a signal to noise ratio of ~ 10-15 given the spectral range of 8 to 14%. Decomposition of the first derivatives of the DSR measurements by R-Mode factor analysis yields a three-factor model that accounts for 97.3% of the downcore variance. These four factors represent 77.9%, 18.0%, and

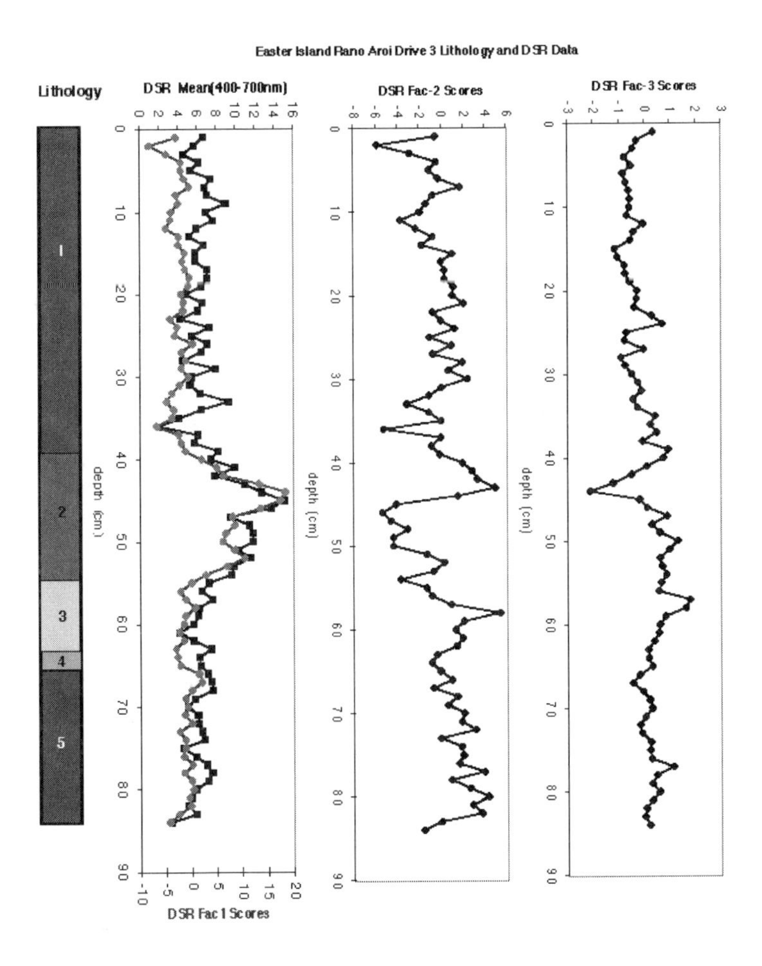

Figure 2. Diffuse spectral reflectance (DSR) measurement (400-700 nm) on the Rano Aroi Drive 3 core, 200-300 cm depth.

1.4% of the DSR variance respectively. While the third factor accounts for a small fraction of the total variance, its inclusion is necessary to explain variance in the wavelength band from 570 – 620 nm. The spectral shapes of the loadings for factors 1 and 2 suggest that they are associated with sediment organic carbon content or redox associated color changes. Factor 3 which exhibits a sharp first derivative peak centered on 590 nm may coincide with a Fe-oxide such as Hematite, which exhibits a spectral peak at 575 nm in marine sediments. The downcore factor scores indicate meter to decimeter scale variations for factor 1, and decimeter to centimeter scale variation for factors 2 and 3.

3.3 Pollen and Spore Analysis

Pollen percentages (Fig. 3) are extremely variable at the 2 cm interval from 284 cm up through 236 cm, which encompasses the dramatic lithologic change from brown organic peat (284 cm – 266 cm) through the grayish organic clay and the return to brown peat at the top of the core. Major pollen types are shown in Figs 4-6. At the base of the drive, from 284-272 cm, the Compositae-Tubuliflorae and Gramineae (grass) are dominant.

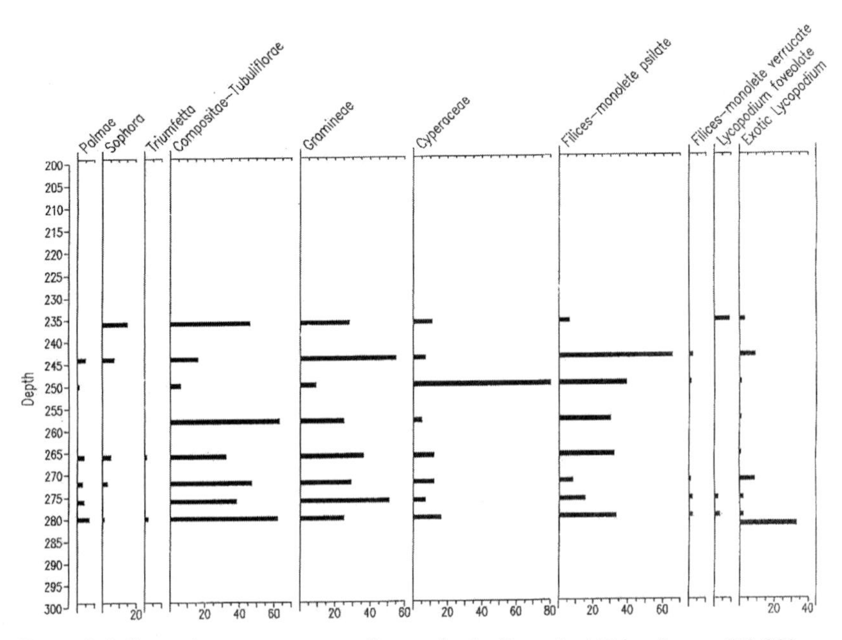

Figure 3. Pollen and spore percentage diagram in the Rano Aroi Drive 3 core, 200-300 cm depth.

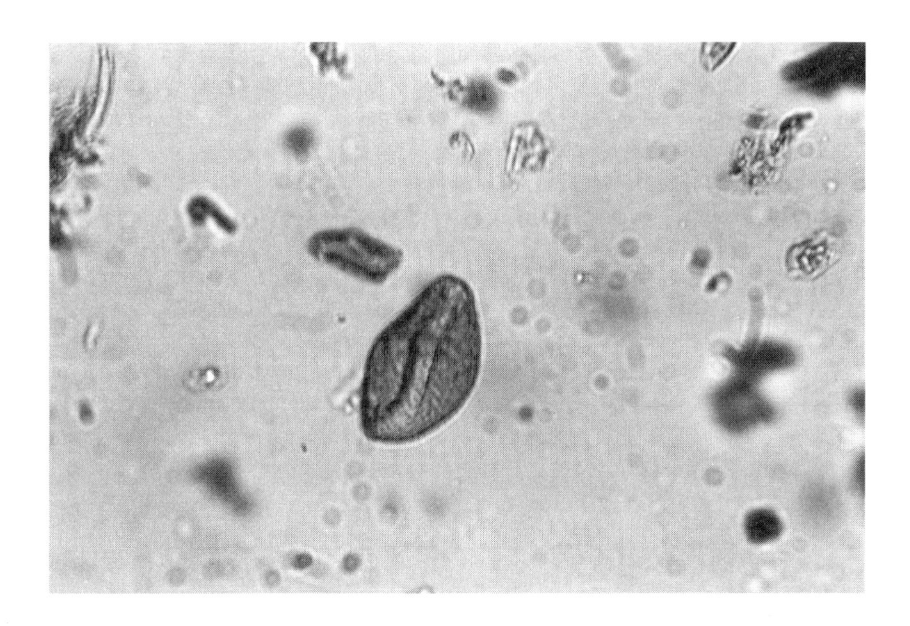

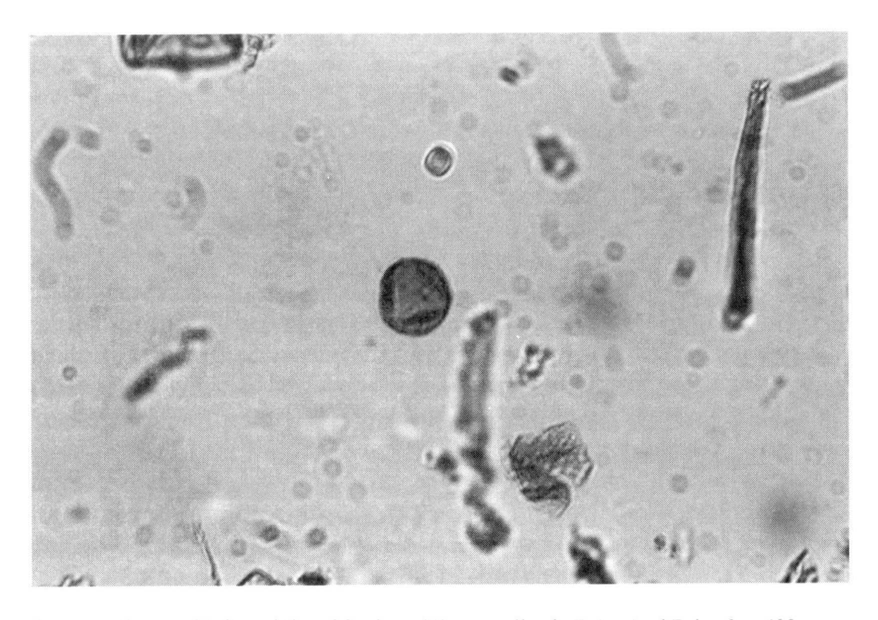

Figure 4. Photos of Palmae (A) and Sophora (B) tree pollen in Rano Aroi Drive 3 at 400x.

Polypodiaceae (monolete) are relatively low, and decline even more in this basal zone. The overlying samples from 266-258 cm show large increases in Polypodiaceae (monolete psilate) spores. They also show relatively large percentages of Compositae–Tubuliflorae and Gramineae. Palmae and Sophora trees are represented in small amounts. At 250 cm, Cyperaceae abundance increases dramatically, and Compositae and Gramineae percentages correspondingly decline. Polypodiaceae spores remain high. The overlying sample at 244 cm shows maximal Polypodiaceae spores, low Compositae-Tubuliflorae, and high Gramineae. The final overlying sample at 236 cm shows extremely low Polypodiaceae spores, but high Compositae-Tubuliflorae and intermediate Gramineae values. Maximal Sophora percentages are present. Exotic Lycopodium percentages show that the total concentrations of pollen and spores are lowest in the basal sediment, increase during the time of inorganic input into the crater, and drop again towards the top of the section.

4. DISCUSSION

The lithologic changes in the Drive 3 of the Rano Aroi cores suggest several major environmental changes at the site during this interval. Use of DSR (Fig. 2) to quantify the visual changes from brown fibrous peat to a grey clay and the return to brown peat show that major geochemical variations are present in the core. The wide disparity in C-14 ages (Fig. 1) shows the difficulty in dating these sediments. However, because of the similarity in late-glacial ages relative to the same depth in Flenley's core, we favor a late-glacial age model for this section at this point.

Pollen and spore analysis shows major fluctuations in pollen and spore percentages throughout this section (Fig. 3). Following Flenley's (1991) climatic interpretation for vegetational groups on the island, we infer Compositae-Tubuliflorae to be indicative of cooler conditions, Gramineae to indicate relatively dry environments, and Polypodiaceae to indicate moist conditions. However, Polypodiaceae increases are also known to be indicative of disturbance, as shown in coastal Alaskan sections after treeline avalanches or volcanic eruptions (Peteet, 1986, Peteet and Mann, 1994). Overall, this section of the core indicates a relatively cool, dry environment with a major disturbance that caused the increase in inorganic sediment increase from about 267-239 cm depth (seen in Fig. 2). The influx of inorganic material is paralleled by major increases in Polypodiaceae, which are probably indications of landscape disturbance. The drop in percentage of exotic Lycopodium at the same time indicates that the total amount of pollen and spores are greater in this interval. This suggests that either more pollen

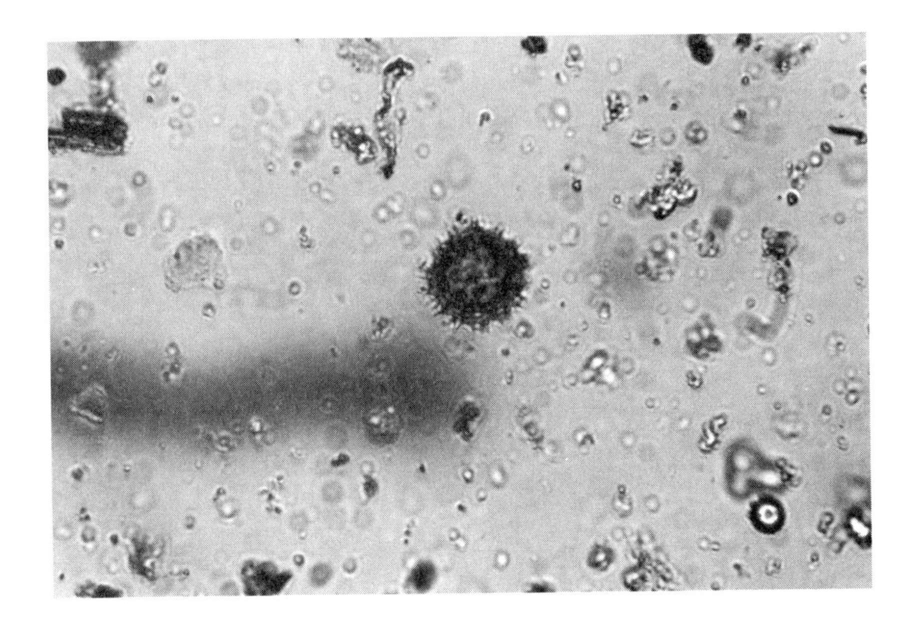

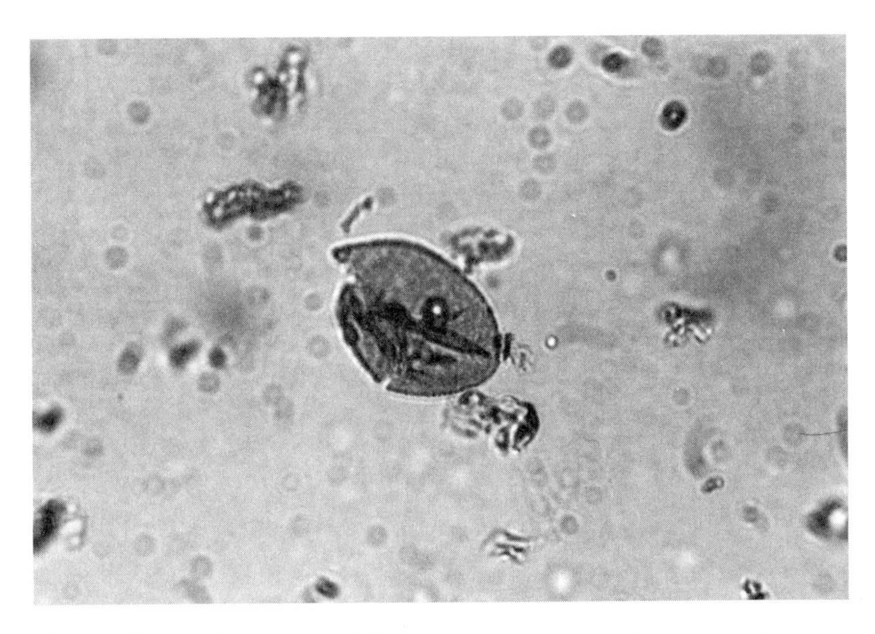

Figure 5. Photos of Compositae – Tubuliflorae (A) and Graminae (B) pollen in Rano Aroi Drive 3 at 400x.

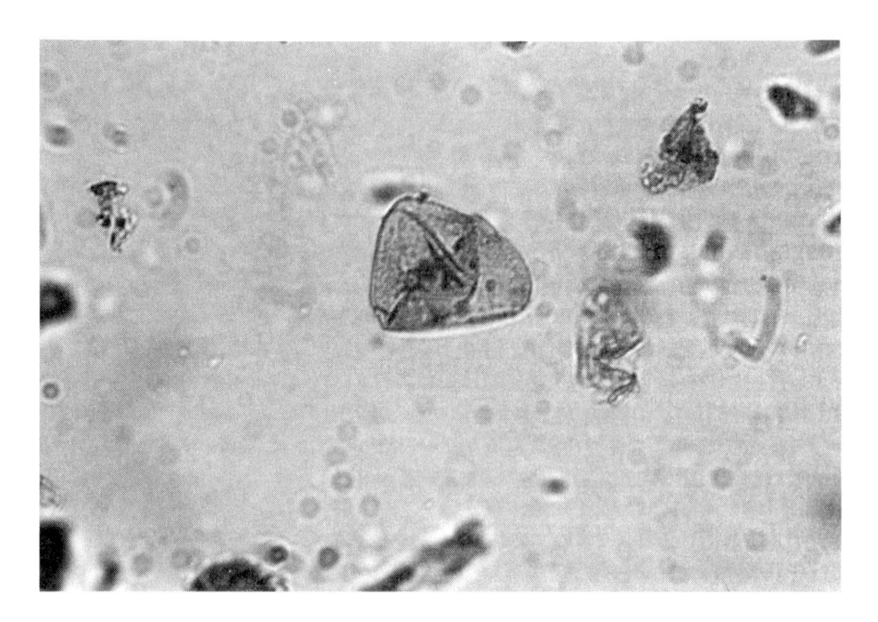

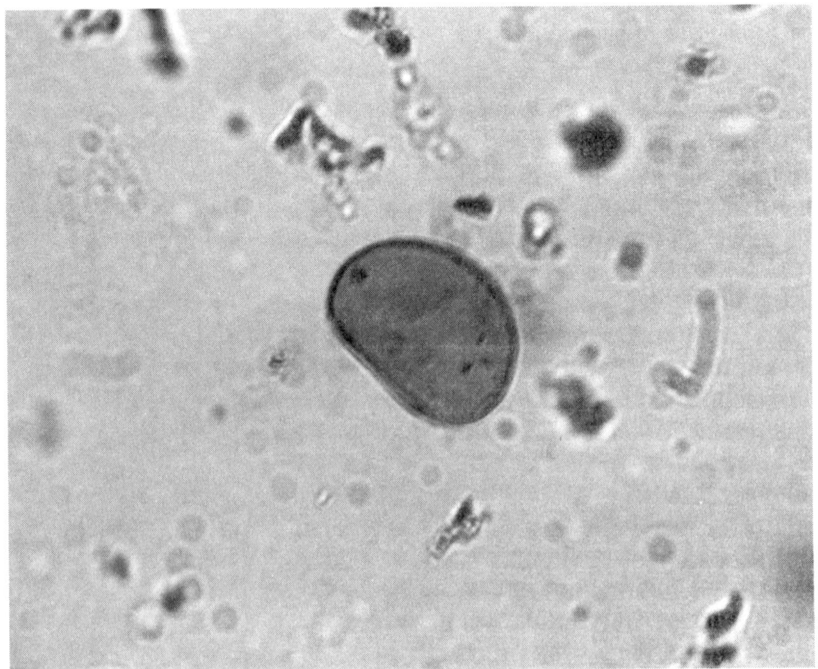

Figure 6. Photos of Cyperaceae (A) pollen and Polypodiaceae (monolete) (B) spores in Rano Aroi Drive 3 at 400x.

is being washed into the crater or more pollen and spores are being produced on the landscape. The sharp increase in Cyperaceae at 250 cm is anomalous, and probably indicates a very wet environment with possible warmer conditions if the decline in Compositae-Tubuliflorae is real and not simply a percentage change artifact. However, the overlying 2 samples show sharp declines in the Cyperaceae and increases in both Compositae-Tubuliflorae and Gramineae. The topmost sample in brown fibrous peat is very low in Polypodiaceae and suggests a cool, dry environment different from any previously.

5. SUMMARY

The reconstructed vegetation during the hypothesized late-glacial to Holocene transition in Rano Aroi is highly variable. Because temperate and subtropical forest throughout the globe experienced extreme changes in vegetation during the late-glacial to Holocene interval, it is not surprising that Easter Island would have also experienced major changes in climate that would have affected the vegetation. However, it is intriguing to examine just how variable the vegetation change can be on a small island. Further 2-cm pollen and spore analysis, macrofossil identification, and AMS dating of non-epidermal material will hopefully clarify the age of these sediments and add to our understanding of the type of extreme variability that this landscape experienced.

REFERENCES

Donahue, D.J., Linick T.W., and Jull, A.J.T., 1990, Isotope-ratio and background corrections for accelerator mass spectrometry radiocarbon measurements. *Radiocarbon* 32(2), 135-142.

Faegri, K., and Iversen, J., 1975, *Textbook of Pollen Analysis*. Hafner, Copenhagen.

Flenley, J.R., and King, S.M., Late Quaternary pollen records from Easter Island. *Nature* 307: 47-50.

Flenley, J.R., King, S.M., Jackson, J., Chew, C., Teller, J.T., and Prentice, M.E., 1991. The Late Quaternary vegetational and climatic history of Easter Island. *Journal of Quaternary Science*. 6:85-115.

Heusser, L.E., and Stock, C.E., 1984, Preparation techniques for concentrating pollen from marine sediments and other sediments with low pollen density. *Palynology*. 8:225-227.

Jull, A.J.T., C. Courtney, D.A. Jeffrey and J.W. Beck, 1998, Isotopic evidence for a terrestrial origin for organic material found in Martian Meteorites ALH 84001 and EETA 79001. *Science* 279, pp. 194-197.

Peteet, D.M., 1986. Vegetational history of the Malaspina Glacier District, Alaska. *Quaternary Research*. 25:100-120.

Peteet, D.M., and Mann, D.H., 1994, Late-glacial vegetational, tephra, and climatic history of southwestern Kodiak Island, Alaska *Ecoscience.* 1(3):255-267.
Wright, H.E. Jr., Mann, D.H., and Glaser, P., 1991. Piston corers for peat and lake sediments. *Ecology.* 65:657-659.

Chapter 6

Mata Ki Te Rangi: Eyes Towards the Heavens

Climate and Radiocarbon Dates

J. WARREN BECK[1], LORI HEWITT[1], GEORGE S. BURR[1], JOHN LORET[2], and FRANCISCO TORRES HOCHSTETTER[3]

[1]*NSF-Arizona AMS Facility, University of Arizona, Tucson, Arizona, 85721;* [2]*Science Museum of Long Island, Plandome, NY 11030;* [3]*Museo Anthropologico Sebastian Englert, Rapa Nui*

1. INTRODUCTION

Among the most enduring enigmas of Easter Island surrounds the giant Moai statues found there. Much has been written about these unusual behemoths, yet why they were made and how they were carried--some distances of over ten miles--is still shrouded in mystery. Fragments of nearly nine hundred such statues were cataloged earlier this century by Father Sebastian Englert (Figure 1), who lived most of his life and wrote extensively on Easter Island ethnology and archaeology (Englert, 1948). Many of these statues were still standing when Europeans first arrived there in the early-18th century, but nearly all were later thrown down and broken during the civil war and cast struggle that broke out around 1840 (Routledge, 1919). Clearly, superior engineering skills and stout materials were required in order to build and move these immense stones, many of which surpass 50 tons in weight and reaching heights of 10 meters. The presence of so many stone monoliths on one small island is indication of a robust and vigorous society. Yet, by the time the earliest European visitors arrived, Rapa Nui civilization was apparently already in significant decline, as no evidence of the engineering skills or materials necessary to move these statues were observed by European visitors. Roggeveen, the captain of the first European ship to visit Easter Island, wrote in his account of these statues:

Kluwer Academic/Plenum Publishers, New York, 2003

> "At first these stone figures caused us to be filled with wonder, for we could not understand how it was possible that people who are destitute of heavy rope or thick timber, and also of stout cordage, out of which to construct gear, had been able to erect them; nevertheless some of these stature were a good 30 feet in height and broad in proportion." (Roggenveen, 1722).

Later, the Englishman Forster, who was amongst the crew of the Cook expedition in 1777, commented similarly about the statues:

> "It was incomprehensible to me how such great masses could be formed by a set of people among whom we saw no tools; or raised and erected by them without machinery" (Forster, 1777).

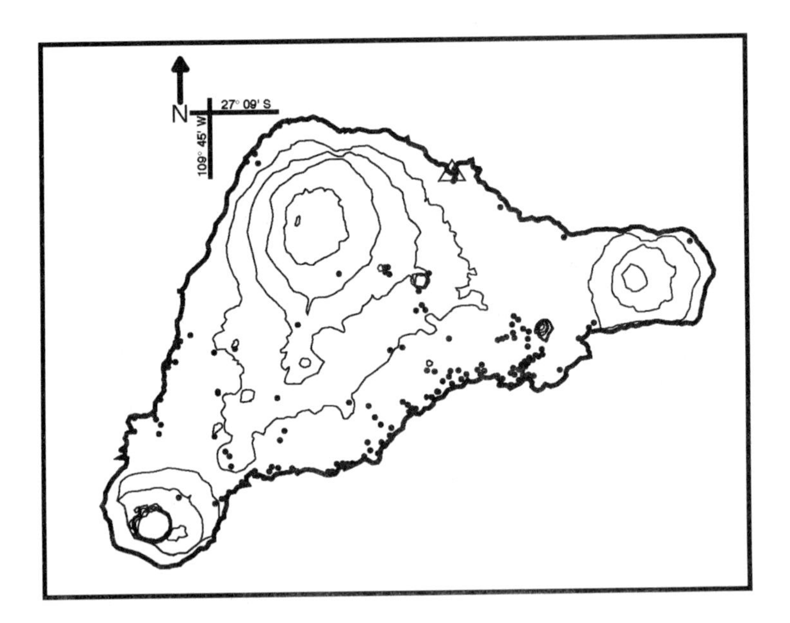

Figure 1. Map of Easter Island showing Locations of known moai statues (After Englert, 1948). Anakena Site is marked with a triangle along the Northern coast.

We now know, in fact, that the Rapa Nui people who carved these statues probably did indeed use ropes and other equipment to help move them as evidenced by the circular holes and the pattern of wear marks surrounding them located near the top of the cliffs above Rano Raraku volcanic crater where most of the statues were quarried. It is generally believed that these once contained capstans and ropes used to lower the moai down the crater walls after being separated from the quarry (Bahn and Flenley, 1992; Mulloy, 1970). Rano Raraku was clearly the hub of moai carving on Easter Island, as over 90% of all moai were carved from the relatively soft grey-brown lapilli tuff forming the crater wall of this extinct volcano (Bahn and Flenley, 1992; Cristino *et al*., 1981; Tilburg, 1986). Even today, many stone carving tools made from relatively hard basaltic lava lie scattered about the quarry site. Dozens of erect moai stand lining both the inside and outside crater walls, and many more can be found in various states of completion in the quarries at the SE corner of the crater. That some of these lie there in a state of virtual completion, give the impression that construction of the statues ended abruptly.

It appears that the faces of the moai were roughed out before removing the statues from the quarries (Bahn and Flenley, 1992). To remove them from the rock wall, the stone carvers would cut moats alone each side using basaltic stone tools, until only a thin keel of rock remained attaching the statue to the quarry wall beneath. The moai must then have been secured with ropes and probably supported underneath with rock wedges while the keel was carefully cut away. This certainly must have been an exceedingly dangerous step in the procedure. Evidence of deaths during moai detachment can, apparently, be found on Rano Raraku today (Bahn and Flenley, 1992).

The Moai statues themselves (Figure 2) are highly stylized, typically with a massive legless torso, long arms and long fingers joined together over the groin area. In most moai a massive head dominates the figure, which commonly has prominent facial planes, a pronounced nose, and usually with very elongated ears. Sometimes, large topknots carved from a red Scoria mined at Punapau quarry several kilometers to the SW of Rano Raraku adorned the moai (Figure 3). Often these topknots were themselves massive, sometimes reached several tons in weight and over 2 meters in height, and were often affixed to the top of the moai using a smaller stone placed in a locking cavity cut into both moai and topknot.

Base relief carvings adorn many moai, most commonly as geometric shapes such as circles or spirals, though some recognizable images of loincloths, birds, Polynesian boats, or more exotic images of unknown significance (Tilburg and Lee, 1987). One moai partially unearthed at Rano Raraku during the Norwegian 1958 expedition to the island, was found to have an image of a three-masted sailing ship carved on it's abdomen (Heyerdahl, 1961). This clearly was a late addition, stimulated by encounters with European culture. It is interesting to note that this statue, and many of the other moai located around the island, also exhibit many

Figure 2. Moai with top knot. Note eye inserts are facing upward.

circular pockmarks called cupuoles on their face and torsos (Tilburg and Lee, 1987). The significance of these cupuoles is not clear. While this may have been part of the engineered program of damage deliberately undertaken by the Rapa Nui people during collapse of their civilization, it is interesting to speculate whether this feature may have been added after exposure to small pox by the European sailors.

Moai can be divided into two classes based on whether or not they had been erected on a raised ceremonial platform called an ahu. In general, those erected on ahus tend to be shorter and broader than those found elsewhere

Figure 3. One of several scoria topknots at the Punapau quarry several kilometers to SW of RanoRaraku. (Author for scale).

(Bahn and Flenley, 1992), and generally have softer facial features. Nearly all moai possess carved eye sockets, though two lacking these were observed by the 1961 Norwegian archaeological expedition (Smith, 1961). One of these was found at the Vinapu site, whereas the other was excavated by Laverchery at the Naunau site in 1954 (Lavachery, 1955) and re-erected by the Heyerdahl expedition in 1955-56. Most commonly, moai faced away from the sea, presumably towards the villages, and the eyes of the statues always pitched their gaze upwards towards the sky, hence the name of this article: Mata ki te Rangi, a Rapa Nui phrase meaning, "eyes towards the heavens".

The question of why these statues were carved is forever lost to history, but the issue of when they were made, is not. The eyes, it turns out, are the key to dating the moai statues. Neither Cook, nor La Perouse observed active stone carving during their visits, and both noted that the Moai appeared at that time to be ancient relics of a past civilization. (Cook, 1777; La Pérouse, 1797). A report made be Geisler in 1883 (Geisler, 1883) based on verbal accounts of citizens living on Rapa Nui at the time, stated that the youngest statues were at the time circa 250 years old, or manufactured during the first half of the 17th century. Nevertheless, moai carving on Easter Island had clearly gone on for some time before this, as evidenced by the use of recycled broken fragments of older moai in the building of new ahu platforms. It is not clear whether this recycling of older moai fragments was merely to make use of available resources, or were deliberately used for some other reason, such as shamanism. In any case, this recycling of older moai into younger ahu makes it clear that there were several generations of ahu and moai building on the island.

According to verbal traditions, Anakena Bay, on the North side of Easter Island is the site at which Hotu Matua and his followers first landed and settled. At least three generations of ahu building have now been recognized at this site, which was substantially restored following archeological investigations during the 1970's and 1980's (Martinsson-Wallin, 1994a). The main ahu at Anakena was designated as "Ahu Anakena" by Lavachery in 1954 (Lavachery, 1955), but was assigned the name ahu Nau Nau in the compilation made by Englert in 1948 (Englert, 1948). Ahu Nau Nau was briefly studied by the Norwegian expedition in 1955-56, and has been studied extensively by archeologists during two major excavations in the late 1970's and again in the late 1980's (Martinsson-Wallin, 1994a; Skjølsvold, 1994). Ten or possibly twelve moai are associated with ahu Nau Nau, though probably no more than eight moai stood on Nau Nau at one time (Rapu, 2001).

Until 1978, the deep-set eye sockets of the moai had always been presumed to be empty. Then, native archaeologist Sonia Haoa made a startling discovery during an excavation at Anakena organized by her fellow archaeologist, Sergio Rapu. Haoa found white coral fragments and a red scoria disk while excavating under a fallen moai, which fit together into the shape of an eye, and then determined that these elements fit snugly together into the eye socket of the overlying moai (Figure 4). Clearly, these inserts were designed to be placed into the eye socket cavities of the moai, giving them an otherworldly appearance dramatically different than the dark brooding countenance they have without them. Bill Mulloy (Mulloy, 1961) had earlier found similar fragments during archaeological excavations at Vinapu on the south coast, but had concluded that these were fragments of a dish manufactured from coral. Since Haoa's discovery many more fragments of coral eye insets have been found at other locations around Easter Island. A total of 57 coral fragments of moai eyes were ultimately discovered during the 1978 excavation of ahu Nau Nau, many of which still bearing tool marks of manufacture (Rapu, 2001). Of these, 52 were collected from underneath four moai that had fallen off the ramp of ahu Nau Nau. Four other fragments were found under the broken head of the only moai to fall off the Northern seaward side of the ahu. These four comprise the nearly complete eye which is currently on exhibit at the Museo Anthropologico Sebastian Englert on Rapa Nui (Rapu, 2001). During the 1986-88 Norwegian led archaeological excavations at Nau Nau a number of additional coral artifacts were discovered as well, though only a few of these were identified as fragments of moai eye inserts.

Figure 4. Sketch of moai and coral eye insert on display at Museo Anthropologico Sebastian Englert, Rapa Nui.

It turns out that the fragments of coral eye inserts can be dated using radiocarbon, giving us another vehicle with which to date moai stone carving practices on Easter Island. Such constraints have proved very difficult to obtain by other means. Because coral contains a large amount of carbon, it is therefore relatively simple to date using ^{14}C. However, there are several caveats, which must be made before interpreting these radiocarbon dates.

First of all, we do not know if the coral eye inserts were carved at the same time as the moai statues, which adds some additional uncertainty to interpretation of these dates. Secondly, we do not know if the coral used to manufacture the eye inserts were collected live, or may have been carved from dead coral utilized long after the death of the coral. Third, even if live corals were collected and immediately used for this purpose, very large head corals are known to live in the reefs surrounding Easter Island. In such cases, the older interior parts of the coral may be significantly older than the living exterior surface recently formed, adding an additional uncertainty term to the interpretation of these radiocarbon dates. In spite of these caveats, however, we feel that the radiocarbon dates obtained from these coral eye inserts provide some of the best available chronological constraints on the timing of moai rock carving practices on Easter Island.

2. MATERIALS AND METHODS

During the 1998 Explorers Club expedition to Easter Island, we subsampled a set of coralline artifacts for radiocarbon analysis from the collection housed at the Museo Anthropologico Sebastian Englert on Easter Island. According to the museum records these were all identified as moai eye fragments, though several of these have subsequently been identified as coral "files" rather than eye fragments recovered during the 1986-88 Norwegian excavations (Martinsson-Wallin, 1998). All of these materials were originally collected during the two archeological excavations of the Anakena site mentioned above, one led by Sergio Rapu in 1978, and the other led by Arne Skjølsvold in 1986-87-88 (Skjølsvold, 1994). Of the samples we analyzed, seven were collected during the 1986-88 Norwegian excavations (Table 1), while fifteen samples were of fragments collected during the 1978 excavation led by Rapu. Eleven of these later samples were reported to be of eye fragments found underneath moai statues that had fallen on the forward ramp side of the Anakena Ahu Nau Nau. According to Rapu (Rapu, 2001), the other four samples were parts of a complete eye that were found by archeologist Haoa in the sand beneath the only broken moai to fall off the seaward side of Ahu Nau Nau. According to museum records, however these four fragments we dated appear to have been found in three different locations at Anakena. Unfortunately, the field notes from the 1978 excavations are no longer available, and the museum records of some of these fragments are sketchy. Consequently, it is difficult to place some of these artifacts in exact archaeological context. According to the museum records, fragments #1388, #1400 and #1407 (Table 1) were recovered from the central west platform of Ahu Nau Nau, whereas fragment #1402 was recovered from the south side of the central platform. Fragments #1396 and #1397 were parts of what was termed complete eyes in the museum records, and were tagged with the additional museum reference "A: 20-30 B: 60-70". Fragment #1587 is tagged with the additional reference: "A: 20-30 B:70-80 P. 05.36". Fragment #2033 was removed from the sand layer at Nau Nau, whereas according to the museum records items #2026 and #2050 were found on the surface at Ahu Nau Nau. Fragment #1947 was removed from underneath moai # 105. There is no additional museum reference information for the other coral fragments dated.

Table 1. Radiocarbon ages for cultural artifacts dated in this study (see Figure 5). Museo#s reference archival numbers of the Museo Anthropologico Sebastian Englert, Rapa Nui, whereas AA#s are archival numbers of the University of Arizona AMS facility. Radiocarbon ages are quoted with 1sigma errors, whereas calibrated age ranges are 2 sigma. Calibrated ages were generated using INTCAL98 (Stuiver *et al.*, 1998) calibration dataset, and CALIB 4.1 software.

SAMPLE TYPE	MUSEO #	AA #	14C age (355yrs) Reservoir Corr.	Calibrated Age (2s)
Coral Eye Whole piece	1402	30996	365±45	1434-1635
Coral Eye Whole piece	2050	30997	365±50	1434-1635
Coral Eye Whole piece	1397	30995	420±45	1408-1627
Coral Eye Fragment	1400	30984	455±45	1327-1494
Coral File	578	31001	505±45	1304-1444
Coral Eye Fragment	2026	30989	525±45	1302-1438
Coral Eye Fragment	A027	31003	610±45	1284-1405
Coral Eye Whole piece	1396	30994	665±50	1223-1264
Coral Eye Fragment	1529	30986	760±45	1159-1292
Coral Eye Fragment	A121	31004	760±45	1159-1292
Coral File	227	31000	770±45	1155-1291
Coral Eye Fragment	5125	30991	780±45	1126-1289
Coral Eye Fragment	2033	30990	810±50	1040-1283
Coral Eye Fragment	1407	30985	815±70	1033-1083
Coral Eye Fragment	5128	30992	830±45	1036-1277
Coral Eye Fragment	1947	30988	875±50	1022-1222
Coral Eye Fragment	1388	30983	885±45	1021-1217
Coral File	A161	30998	930±45	997-1206
Coral File	785	31002	940±45	992-1205
Coral Eye Fragment	1587	30987	960±45	982-1188
Coral File	A161	30998	975±60	897-1192
Coral File	A160	30999	985±65	890-1206
Coral Eye Fragment	5129	30993	1290±45	618-804
Algal Nodule 1		27343	371±50	1430-1635
Algal Nodule 2		27344	423±60	1331-1632
Algal Nodule 3		27345	555±45	1298-1432
Algal Nodule 4		27346	430±40	1402-1617

In addition to the coral eye fragments, four white calcareous algal nodules decorating the plaza in front of the ramp were also collected during the 1997 Science Museum/Explorers Club sponsored expedition to Easter Island. These were dated, in order to give an independent age estimate of the age of the plaza (Table 1). Three of these (nodules #1, 2, & 4) were collected 3-5 m in front of the lower edge of the Ahu Nau Nau ramp, near the center of the ramp. The other nodule (#3) was collected approximately 10 m south of the east corner of the reconstructed Ahu Nau Nau ramp. These nodules were integrated into the pavement on the plaza, possibly as part of a decorative motif distributed between the basaltic boulder paving stones and appeared to have been deliberately placed in the plaza. These nodules are shallow marine in origin, and grow very slowly by secretion of thin concentric shells made of $CaCO_3$. The algae live on the outer surface of the nodule, and periodically deposit new layers of calcium carbonate from material that the algae extracts from seawater. Thus, the nodules are oldest at their centers and youngest at the outer surface. Consequently, only the outer-most layers of these nodules were sampled so that the ages would most closely represent the timing the nodules were removed from the ocean, and presumably, placed on the plaza.

For the coralline artifacts sampled at the museum, small samples were drilled from these coralline artifacts using a Dremel Tool with typical sample sizes of 20 mg, yielding approximately 2.5 mg of carbon per sample. These samples were processed by vacuum hydrolysis for radiocarbon measurements at the University of Arizona accelerator mass spectrometry (AMS) facility in Tucson, Arizona (Donahue *et al.*, 1990).

Corals living in the sea around Easter Island today make their skeletons from carbon dissolved in the ocean surface water. Carbon in the surface water, however, does not have a zero age, because a small fraction of that carbon is derived from older deeper ocean water. This old carbon generates what is called a "reservoir effect", which for most surface waters of the equatorial Pacific makes coral radiocarbon ages about 400 years older than the true age (Bard *et al.*, 1989). Since this reservoir effect is somewhat variable in time and space, it must be determined locally in order to subtract the effect from the radiocarbon ages before they can be used for archeological purposes. During the August 1997 Science Museum/Explorers Club sponsored expedition, we collected drill cores from living corals from the reefs around Easter Island using SCUBA equipment and an underwater drilling apparatus (Mucciarone and Dunbar, 2001). These cores were subsequently slabbed with a diamond circular saw and x-radiographed at the University of Arizona Medical Center which allowed us to count the annual density bands back to a time period to 1955 AD. We then performed a series of radiocarbon age measurements on pre-1955 coral material of known age from one of these cores (Table 2). This allowed us to determine the surface ocean reservoir effect for Easter Island. Coral material older than 1955 had to be used because atmospheric testing of thermo-nuclear weapons in the late 1950's and 1960's generated large amounts of ^{14}C in the atmosphere. This so-called "bomb pulse" of radiocarbon has subsequently been adsorbed by the ocean which has caused a significant change in the surface ocean reservoir effect for times after 1955. Based on our measurements, the mean pre-bomb-pulse surface ocean reservoir effect for Easter Island is 355±18 years (1 σ_m). This reservoir effect has been subtracted from all of our radiocarbon age measurements made on archaeological artifacts we

Table 2. Data used for calculation of pre-bomb-pulse modern marine reservoir correction at Easter Island. These measurements were made on drill cores recovered from live corals collected (See Mucciarone and Dunbar, 2001) during the 1997 Easter Island Expedition sponsored by the Science Museum of Long Island, and the Explorer's Club. Fraction Modern values are reported with 1 sigma errors.

AA#	Sample Name	Fraction Modern	^{14}C Age
AA40624	OVAHE 97-1 706mm	0.9556±0.0060	365
AA40773	OVAHE 97-1 706-709mm	0.9624±0.0047	308
AA41056	OVAHE 97-1 710mm (1)	0.9602±0.0056	326
AA41056	OVAHE 97-1 710mm (2)	0.9577±0.0042	347
AA41056	OVAHE 97-1 710mm (3)	0.9586±0.0034	340
AA40774	OVAHE 97-1 710-714mm	0.9610±0.0068	320
AA41057	OVAHE 97-1 722-723mm	0.9450±0.0043	454
AA41066	OVAHE 97-1 724mm	0.9493±0.0042	418
AA42280	OVAHE 97-1 729-735mm	0.9499±0.0043	413
AA41067	OVAHE 97-1 732-733mm	0.9572±0.0054	351
AA42287	OVAHE 97-1 736-741mm	0.9492±0.0043	419
AA42613	OVAHE 97-1 742-749mm	0.9431±0.0043	471
AA42278	OVAHE 97-1 764-756mm	0.9548±0.0044	371
AA42285	OVAHE 97-1 774-765mm	0.9519±0.0043	396
AA42611	OVAHE 97-1 784-775mm	0.9501±0.0043	411
AA40626	OVAHE 97-1 796-792mm	0.9520±0.0047	395
AA40766	OVAHE 97-1 808-798mm	0.9786±0.0046	174
AA40765	OVAHE 97-1 809-816mm	0.9658±0.0050	280
AA40767	OVAHE 97-1 817-824mm	0.9720±0.0046	228
AA40768	OVAHE 97-1 825-831mm	0.9564±0.0052	358
AA40769	OVAHE 97-1 832-840mm	0.9679±0.0045	262
AA40517	OVAHE 97-1 856-863mm	0.9584±0.0049	341
AA40771	OVAHE 97-1 864-867mm	0.9505±0.0043	408
	Average ^{14}C Reservoir Age		355
	Standard deviation		71
	Standard Error		18

measured which were derived from the ocean (Table 1), and was assumed to be constant for all time prior to the bomb pulse period. An additional uncertainty term of 18 years has also been added quadratically to the analytical error of the radiocarbon measurement to account for the variability in the reservoir effect (Donahue *et al.*, 1990).

As is well known, a radiocarbon age is generally not equivalent to a calendar age, because the concentration of ^{14}C in the atmosphere (or the ocean surface) has varied through time. It is, however, possible to derive a calendar age from the radiocarbon age using a calibration based on radiocarbon measurements of tree rings. In our case, calendar ages were derived from marine-reservoir-corrected radiocarbon ages using the INTCAL98 radiocarbon calibration dataset (Stuiver *et al.*, 1998). Both radiocarbon age and calendar ages are presented in Table 1. Calibrated age ranges are also presented in Figure (5). Radiocarbon ages are reported with one standard deviation error, whereas calibrated ages are reported as two standard deviation error envelopes, and include error terms for measurement error as well as uncertainty in the reservoir effect.

3. INTERPRETATION

As mentioned above, all of the artifacts dated in this study are reported to be from Ahu Nau Nau at Anakena. The 1978 and 1986-88 excavations at Anakena revealed that there were at least three periods of construction at this ahu (Martinsson-Wallin, 1994a; Skjølsvold, 1994). The oldest construction phase, termed Nau Nau I, is believed to dated from c. 1100 AD. Nau Nau III, which was largely reconstructed by Rapu and colleagues in 1978, is thought to date from c. 1300-1410 AD, whereas an intermediate ahu building phase (Nau Nau II) probably dates from c. 1190-1380 AD (Martinsson-Wallin, 1994a). There is evidence, however, that the initial settlement of the Anakena site may have predated ahu building by 200-300 years (Martinsson-Wallin, 1994a).

Figure 5 presents calibrated radiocarbon ages (Stuiver *et al.*, 1998) for all the marine–reservoir corrected radiocarbon ages of the coral and algal artifacts dated in this study. These are plotted as both one and two standard deviation age ranges. The dates range from 680-770 AD (1σ) for the oldest coral fragment (Museo# 5129; AA# 30993), to 1440-1620 (1σ) for the youngest coral eye fragments (Museo# 1402; AA#30996). The oldest

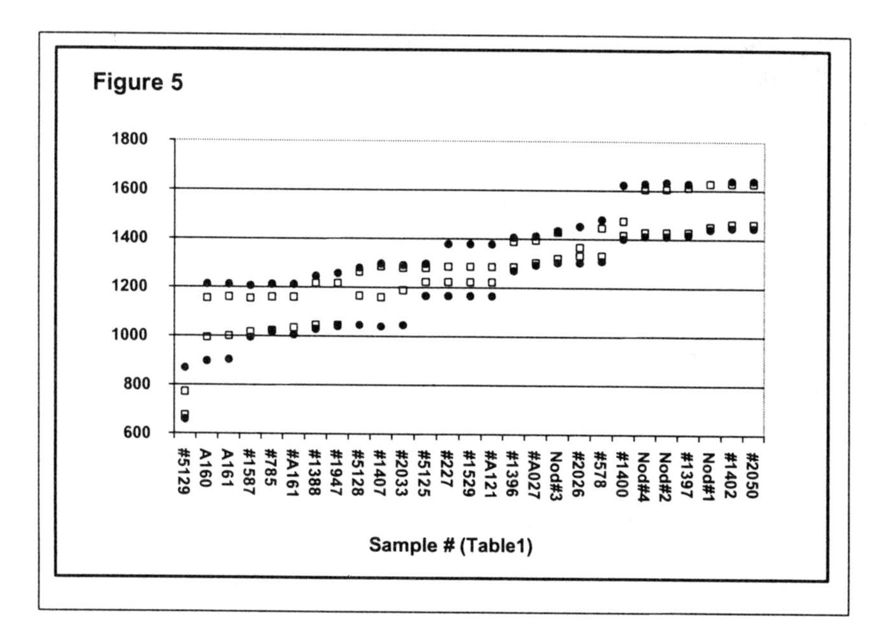

Figure 5. Calibrated radiocarbon age (years AD) ranges for cultural artifacts dated in this study (see Table 1). Calibrated ages were generated using INTCAL98 (Stuiver *et al.*, 1998) calibration dataset, and CALIB 4.1 software. Upper and lower open squares (and upper and lower solid circles) represent 1 sigma (and 2 sigma) bounds on calibrated ages for each sample. Sample #s are the same as used in Table 1.

age is significantly older than any other artifacts found at Anakena, and only two other radiocarbon ages from cultural horizons on Easter Island are in this time range. One of these (sample K-501) is a charcoal recovered from the bottom of Poike Ditch (Smith, 1961), which gave a calibrated radiocarbon age range of 410-600 AD (1 σ). The other cultural artifact with an old radiocarbon age is from unburned Tortora reeds found in a grave site at Ahu Tepeu, (M-732) (Smith, 1961). This sample has large analytical errors, and gives a calibrated radiocarbon age of 130 BC-645 AD (1σ). Both of these prior radiocarbon dates have been criticized because they are so much older than any other reliable cultural dates on Easter Island, and if they were to be correct, there would appear to be a significant epoch of cultural history that is missing from the archeological record. For this reason, we made a second analysis of coral fragment #5129, subjecting it to an additional acid partial dissolution pretreatment step prior to hydrolysis to determine if the original date may have been skewed by contamination with ancient soil carbonate. This second analysis yielded the same result, however. Thus, this coral artifact yields a radiocarbon age that is amongst the oldest of any reported cultural artifact on Easter Island. Unfortunately, almost no information about the archeological context of coral fragment #5129 is available from either the Museum records, or from the available records of the 1978 excavation that unearthed the artifact. Thus, based on the limited information about this fragment, we can draw only very weak conclusions about the significance of this artifact. This age is supportive of early colonization of Easter Island at circa 700 AD, but it is certainly possible that this object is not really a cultural artifact.

It is interesting to note that the calibrated radiocarbon ages shown in Figure 5 appear to fall into several discrete clusters of ages. The oldest age discussed in the previous paragraph, falls by itself, but the rest of the ages appear clustered in three additional groups centered around 1100 AD, 1300-1400 AD, and 1450-1620 AD. Several potential explanations for this clustering come to mind. On the one hand, the clustering may be purely coincidental and stem from the relatively small number of radiocarbon ages in the population. Alternatively, the clustering may be an artifact of the structure of the calibration curve in this region of time. A more interesting possibility is that this clustering may represent three separate phases of ahu construction in which these marine artifacts were used as decorations. Along this line of reasoning, the oldest population of dates is roughly coincident with the timing of development of Nau Nau I (Martinsson-Wallin, 1994a; Skjølsvold, 1994), while the second population roughly coincident with the timing of the building of Nau Nau III (Martinsson-Wallin, 1994a). Based on the verbal accounts recorded by Geiseler in 1883 (Geisler, 1883) the third population of dates appears to coincide with the last phase of statue manufacturing in Rano Raraku, which ended around 1630 AD. Three of the four algal nodules removed from the Ahu Nau Nau III plaza reconstructed by the Rapu excavation in 1978, fall in this last cluster of dates (Table 1). This suggests that this plaza, or at least this phase of decoration of the plaza is a relatively young feature at Anakena.

An old phase of ahu construction was discovered during the 1986 Norwegian excavations of trenches C1 and C2 (Skjølsvold, 1994), which were expanded from Rapu's earlier 1978 Trench A excavation. This structure is considered to be from the primary settlement period of Anakena, termed Nau Nau I ((Skjølsvold, 1994)), though a wing of this structure may also by linked to the later Nau Nau II phase of construction (Martinsson-Wallin, 1998). Five artifacts from this horizon were dated by the Norwegian group to constrain the timing (Skjølsvold, 1994) of Nau Nau I development. The youngest of these, (T-7959) which gave a radiocarbon age of 510±40 BP, was rejected because it would make this cultural horizon younger than layers above it (Skjølsvold, 1994). Sample Ua-1740 (1290±100 BP) is from a bone of a sea bird (Skjølsvold, 1994) which should require a marine reservoir correction because it's diet was derived from marine organisms. Using the same marine correction we determined (Table 2) this would make this age (935±100), which is in agreement with the other radiocarbon dates obtained from this cultural Horizon (T-6679, 1170±140 BP; T7341, 900±120 BP; Ua-3007, 1015±65 BP) (Skjølsvold, 1994).

The coral artifact A-161 (AA#30998), was also recovered from the bottom of trench C1 during the 1986-88 Norwegian excavations, and was identified as a 5.2 cm long triangular type file (Skjølsvold, 1994). Such files are common artifacts recovered from Polynesian excavations in the Marqueses, and are thought to have been used for grinding and shaping other tools such as fish hooks from shell material(Skjølsvold, 1994). We dated this artifact twice, giving marine-reservoir corrected radiocarbon ages of 930±45 BP and 965±60 BP, with a weighted average age of 981±36 BP. This age is consistent with the other radiocarbon ages measured by the Norwegian group for development of Nau Nau I.

Four other coral artifacts were also dated which were originally identified as moai eye fragments in museum records, but which have subsequently been identified as four of the 94 coral files collected during the Norwegian 1986-88 excavations of Anakena. The items we dated are museum artifacts #A-160 (AA#30999), #B227 (AA#31000), #B578 (AA#31001), and #B785 (AA#31002). Like artifact A-161, item A-160 is associated with the early Nau Nau I habitation phase of Anakena and was also found in the bottom layer of trench C1 (Skjølsvold, 1994). It is a teardrop-shaped file yielding a radiocarbon age of 1020±65 BP (after marine-reservoir correction), very similar in age to artifact A-161.

The other three coral files which were dated here are from the Nau Nau East (Martinsson-Wallin, 1994b) site at Anakena, which is located about 90 m to the east of SE corner of the ramp of Nau Nau III. This area was settled until about 100 years ago, and exhibits evidence of extended periods of habitation. Many features were found here during the Norwegian excavations including numerous fire pits, hearths, scattered stone images, and many cultural tools. Two radiocarbon dates of 810 ±80 BP (T-7345) and 810±70 BP (T-7346) were derived for charcoals from the basal cultural layers of Nau Nau East by the 1986-88 Norwegian expedition (Skjølsvold, 1994). During these excavations, fourty-three files were recovered from the cultural layer at Nau Nau East, thirty-six of which were coral. One of these, Museo# B227 (AA#31000), is a long triangular type file with almond shape, found 60-80 cm beneath the surface (Martinsson-Wallin, 1994b). It gave a reservoir-corrected radiocarbon age of 770±45 years BP, very similar in age to the charcoal ages from the basal cultural layer of Nau Nau East. The other coral file we dated, Museo #B578 (AA#31001), was also recovered from the same cultural layer at Nau Nau East, and is described as short triangular type file, but was recovered from a slightly shallower depth of 40-60 cm beneath the surface (Martinsson-Wallin, 1994b). This artifact gave a somewhat younger reservoir-corrected radiocarbon age of 505±45 BP. The last coral file dated here is Museo#B785 (AA#31002), which also was collected from Nau Nau East and is described as a rectangular or long triangular type fragment with one convex side and a groove in the middle (Martinsson-Wallin, 1994b). It was also found in the cultural horizon, but at a depth of 60-80 cm below the surface. This sample yielded a reservoir-corrected radiocarbon age of 940±45 BP, which is significantly older than the other estimates of the basal age of this cultural layer, suggesting that it may have been inhabited coevally with Nau Nau I.

Seventeen of the coral artifacts we radiocarbon dated in this study (Table 1) are listed in Museo Anthropologico Sebastian Englert records as being fragments of moai eye inserts. Fifteen of these were recovered during the 1978 excavations of Anakena led by Rapu, the other two (Museo#A121 and #A027) were recovered during the 1986-88 Norwegian excavations of Anakena. Sample A121 (AA#31004) was recovered from the upper layer of trench K on the seaward side of Nau Nau III, giving a marine reservoir-corrected radiocarbon age of 760±45 BP. Sample A027 (AA#31003) was recovered from the upper cultural layer of trench E, also on the Seaward Side of Nau Nau III. Sample A027 yielded a marine reservoir-corrected radiocarbon age of 610±45 BP. These ages are both consistent with ages of other artifacts recovered from Trench K (T-7350, top layer, 710±80BP; T-7974 bottom layer, 540±60 BP), and Trench E (T-7343, bottom cultural layer, 750±100 BP; T-7344, upper cultural layer, 600±140 BP)(Skjølsvold, 1994).

Of the samples collected during the Rapu excavation, samples Museo#1402 (AA#30996), Museo#2050 (AA#30997), Museo#1397 (AA#30995), and Museo#1400 (AA#30984) all cluster in the youngest population of calibrated ages (c. 1450-1600 AD) similar to the algal nodules, which as mentioned above, roughly correlates with the last period of moai statue building at Rano Raraku (Geisler, 1883). Of these, samples #1402 and #1400 were found on the south side and west side of the Nau Nau III respectively. No other information is available concerning these samples.

A second population of moai eye fragments cluster around an intermediate calibrated age range of 1300-1400 AD, which roughly coincides with the age of Nau Nau III. This population includes samples Museo#A121 and Museo #A027 mentioned above, as well as Museo #2026 (AA#30989), Museo #1396 (AA#30994), Museo #1529 (AA#30986) and Museo #5125 (AA#30991). The two mentioned above were found on the seaward side of Nau Nau III, but no additional information is available on the other four other than they were collected at Anakena in 1978.

A third population of moai eye fragments formulate the oldest cluster centered around 1100 AD (calibrated radiocarbon ages), which apparently correlates with Nau Nau I construction. These samples are Museo #2033 (AA#30990), Museo #1407 (AA#30985), Museo #5128 (AA#30992), Museo #1947 (AA#30988), Museo #1388 (AA#30983), and Museo # 1587 (AA#30987). Two of these (#1388 and #1407) were found near the central west platform of Nau Nau III. Sample #2033 was found in a sandy layer at Anakena, and sample #1947 was found beneath Moai #105 (Using Englert's numbering system (Englert, 1948)) at Nau Nau III. Little is known about the other samples other than they are thought to be moai eye fragments from Anakena recovered by the Rapu 1978 excavations. It is interesting to note that this oldest population of dates is apparently coincident with construction of Nau Nau I phase of construction at Anakena. This indicates that use of coral for moai eye inserts appears to have evolved at the same time as construction of the Nau Nau I, one of the oldest ahus on Easter Island. This suggests that the practice of moai construction (and the use of coral to manufacture eye inserts) evolved simultaneously with the practice of ahu construction.

4. CONCLUSIONS

Twenty seven cultural artifacts from Easter Island were dated in this study by AMS^{14}C at the University of Arizona. These artifacts were all of marine origin (Corals or algal nodules) which required establishment of a value for the local marine reservoir effect. We established this reservoir effect at 355 ±18 years (1 σ_m), based on ^{14}C measurements of pre-1955 age corals, which were collected using SCUBA in 1997 during the Explorers Club sponsored expedition to Easter Island.

Three classes of cultural artifacts were dated in this study, including fragments of moai eye inserts, or coral files, which were recovered during the 1978 excavations of Anakena led by Sergio Rapu, and also during the 1986-88 Norwegian excavations led by Arne Skjølsvold. The third class of cultural artifacts dated consisted of decorative algal nodules collected from the plaza in front of Nau Nau III at Anakena, which were collected during the 1997 Explorers Club expedition to Easter Island. The calibrated radiocarbon ages from these artifacts agree well with previously determined charcoal, leaf and bone radiocarbon dates from equivalent archaeological strata. Our new dates fall into four populations. The oldest (680-770 AD (1σ)) falls by itself, and is among the oldest radiocarbon dates for any cultural artifacts previously dated on Easter Island. This artifact is of dubious provenance, however, and thus is not certain to be a cultural artifact.

The remaining ages appear to cluster in three additional groups centered around 1100 AD, 1300-1400 AD, and 1450-1620 AD. These three clusters appear to roughly coincide with the timing of development of Nau Nau I (Martinsson-Wallin, 1994a; Skjølsvold, 1994), the time of the building of Nau Nau III (Martinsson-Wallin, 1994a), and thirdly, with the last phase of statue manufacturing in Rano Raraku, which ended around 1630 AD (Geisler, 1883). Three of the four algal nodules removed from the Ahu Nau Nau III plaza reconstructed by the Rapu excavation in 1978, fall in this last cluster of dates, suggesting that this plaza, or at least this phase of decoration of the plaza is a relatively young feature at Anakena. The oldest population of coral moai eye insert fragments appears to coincide with construction of the Nau Nau I phase of construction at Anakena. This indicates that use of coral for moai eye inserts may have evolved at the same time as construction of the Nau Nau I, and suggests that the practice of moai construction (and the use of coral to manufacture eye inserts) evolved simultaneously with the practice of ahu construction.

ACKNOWLEDGMENTS

This work was supported by the Science Museum of Long Island and the National Science Foundation, and was conducted in collaboration with the Explorers Club. We thank the people of Easter Island and the Chilean Consejo de Monumentos Nacionales for permission to conduct this research, and Sergio Rapu and Helene Martinsson-Wallin for their insightful assistance in facilitating this research.

REFERENCES

Bahn P., and Flenley J., 1992, *Easter Island Earth Island.* Thames and Hudson, Ltd.

Bard E., Labeyrie L., Arnold M., Labracherie M., Pichon J.-J., Duprat J., and Duplessy J.-C., 1989, AMS-^{14}C Ages Measured in Deep Sea Cores from the Southern Ocean: Implications for Sedimentation Rates during Isotope Stage 2. *Quaternary research* **31**, 309-317.

Cook J., 1777, *A Voyage towards the South Pole, and round the World. Performed in His Majesty's ships resolution and Adventure. 2 Volumes.*

Cristino C., Casanova P. V., and Izaurienta y. R., 1981, Atlas Arqueologica de la Isla de Pascua. Facultad de Arquitectura y Urbanismo, Instituto de Estudios, Universidad de Chile.

Donahue D. J., Linick T. W., and Jull A. J. T., 1990, Isotope-ratio and background corrections for accelerator mass spectrometry radiocarbon measurements. *Radiocarbon* **32**(2), 135-142.

Englert F. S., 1948, la Tierra de Hotu Matu' a: Historia, Etnologia y Lengua de la Isla Pascua. Padre Las Casa (Chile): "San Francisco".

Forster G., 1777, A Voyage round the World in His Britannic Majesty's Sloop. Resolution, Commanded by James Cook, during the Years 1772-1775.

Geisler K., 1883, Die Osterinel, eine Stätte prähistorischer Kultur der grossen in der Südsee. In *Bieheft aum Marine-Verordnungsblatt*, pp. 1-54.

Heyerdahl T., 1961, Surface Artifacts. In *Reports of the Norweigian Archaeoligical Expediton to Easter Island and the East Pacific. Monograph of the school of Americal research and the Museum of New Mexico.*, Vol. 1 (ed. T. Heyerdahl and E. N. Ferdon).

La Pérouse J. F. d. G., Comte de. 1797, *Voyage de La Pérouse autour de monde.*

Lavachery H., 1955, Archeologie de l'Ile de Paques. le Site d'Anakena. *Journ. Soc. des Océanists* **10**(10), 133-158.

Martinsson-Wallin H., 1994a, The Ceremonial Stone Structures of Easter Island. Analysis of variations and Interpretations of Meanings. PhD., Uppsala University.

Martinsson-Wallin H., 1994b, The Settlement/Activity Area Nau Nau East at Anakena, Easter Island. In *The Kon-Tiki Museum Occasional Papers : Archeological Investigations at Anakena, Easter Island.*, Vol. #3 (ed. A. Skjølsvold), pp. 123-200. The Kon-Tiki Museum.

Martinsson-Wallin H., 1998, Personal Communication.

Mucciarone D. A., and Dunbar R. B., 2001, Stable Isotope Record of El Niño-Southern Oscillation Events from Easter Island (ed. J. Loret).

Mulloy W., 1961, The Ceremonial Center of Vinapu. In *Reports of the Norwegian Archaeological Expedition to Easter Island and the East Pacific. Monograph of the School of American Research and the Kon-Tiki Museum.*, Vol. Vol. 1 (ed. T. Heyerdahl and E. R. Ferdon), pp. 93-180. The Kon-Tiki Museum.

Mulloy W., 1970, A speculative reconstruction of techniques of carving, transporting and errecting Easter Island Statues. *Archaeology and Physical Anthropology in Oceania* **5**(1), 1-23.

Rapu S., 2001 Personal communication.

Routledge K., 1919, *The Mystery of Easter Island.* Hazell, Watson, and Viney.

Skjølsvold A., 1994, Archeological Investigations at Anakena, Easter Island. In *The Kon-Tiki Museum Ocassional Papers*, Vol. Vol.3. The Kon-Tiki Museum.

Smith C. S., 1961, A Temporal Sequence derived From Certain Ahu. In *Reports of the Norweigian Archaeological Expedition to Easter Island and the Eastern Pacific.*, Vol. 1 (ed. T. Heyerdahl and E. N. Ferdon). Monograph of the School of American Research and the Museum of new Mexico, #24.

Stuiver M., Reimer P., Bard E., Beck J. W., Burr G. S., Hughen K., Kromer B., McCormac G., Plicht J., v. d., and Spurk M., 1998, INTCAL98 Radiocarbon Age Calibration, 24,000-0 cal BP. *Radiocarbon* **40**(#3), 1041-1083.

Tilburg J. A. V., 1986, Power and Symbol: The stylistic analysis of Easter Island monolithic sculpture. PhD, UCLA.

Tilburg J. A. V., and Lee G., 1987, Symbolic stratigraphy: Rock art and the monolithic statues of Easter Island. *World Archeology* **19**(2), 134-149.

Chapter 7

Stable Isotope Record of El Niño-Southern Oscillation Events from Easter Island

DAVID A. MUCCIARONE AND ROBERT B. DUNBAR
Department of Geological and Environmental Sciences, Stanford University, Stanford, California, 94305

1. INTRODUCTION

Easter Island (also known as Rapa Nui and Isla Pascua) lies within the southeastern Pacific high-pressure system, a feature that along with the Indonesian Low comprises the atmospheric dipole that defines the Southern Oscillation. Sea surface temperatures (SST) in the southeastern Pacific influence this limb of the basin-wide Walker circulation by modulating the stability and magnitude of convection within the regionally descending air. El Niño-Southern Oscillation (ENSO) research has most often focused on variability in the intensity and location of the Indonesian Low convective system or on teleconnections to various parts of the Northern Hemisphere. Long climate records from Easter Island will help elucidate the influence of oceanic variability on the overall ENSO system and its South Pacific teleconnections via the Walker Circulation. In addition, the Easter Island region of the South Pacific Gyre is a source for the shallow subsurface meridional flow that eventually upwells along the equator in the central and eastern Pacific [(Fig. 1); Levitus, 1982; Ji, *et al.*, 1995; Gu and Philander, 1997]. In the northern Pacific, subsurface meridional flow has been suggested as a cause of decade-scale climate anomalies (Gu and Philander, 1997; Zhang *et al.*, 1998). A similar mechanism may operate in the Southern Hemisphere; however, our current lack of a long time series of oceanic climate data from the eastern South Pacific Gyre, limits our ability to study

Easter Island, Edited by John Loret and John T. Tanacredi

Kluwer Academic/Plenum Publishers, New York, 2003

this phenomenon. Proxy records of corals from a gyre-central site like Easter Island can significantly extend the limited instrumental data presently available.

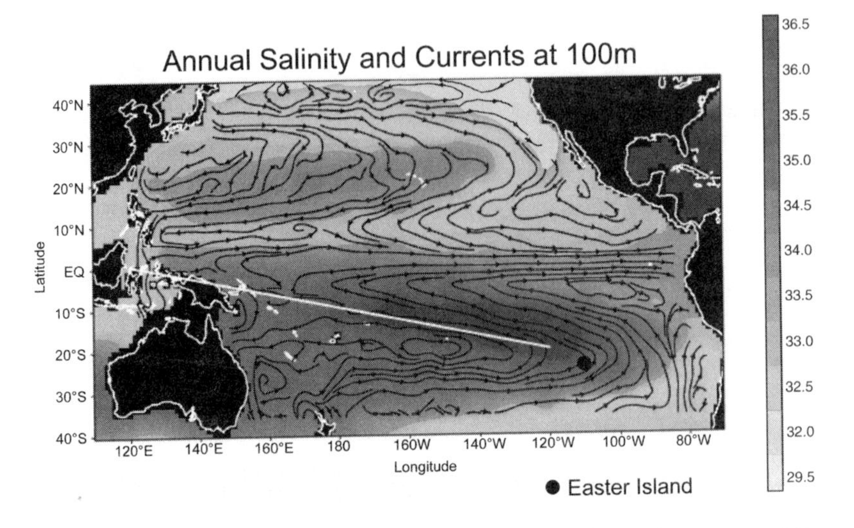

Figure 1: Mean annual salinity at 100m from Levitus (1982). The overlay arrows and lines show streamlines of the 1989 mean flow at 100 m from the analysis of Ji (1995). Note that streamlines at 100 m transport water from the location of Easter Island (27°S, 109°W) to a position west of the dateline and then eastward to the eastern Pacific cool tongue region. This is one possible mechanism by which thermal anomalies generated in the SE Pacific may influence tropical temperatures over decadal time scales (Gu and Philander, 1997).

Much of our current knowledge of interannual and decadal climate variability derives from the extensive data sets from the Northern Hemisphere. The Pacific-North American pattern (PNA), Pacific Decadal Oscillation (PDO), and North Atlantic Oscillation (NAO) were among the first non-ENSO climate patterns to be discovered. These new insights were obtained because regional temperature, wind, and pressure data sets extend back many decades, sufficiently far in time that interannual and decadal patterns reveal distinct patterns of climate. Ocean-atmosphere data sets of similar length are unavailable for most of the Southern Hemisphere and our understanding of large-scale coherent climate patterns in the southern tropics and subtropics lags that of the Northern Hemisphere.

Nevertheless, several coherent climate patterns have been proposed for the South Pacific. White and Peterson (1996) describe a circumpolar interaction between wind, sea ice, sea level pressure (SLP), and sea surface temperature (SST) that propagates eastward over interannual time scales as the Antarctic Circumpolar Wave (ACW). A Pacific-South American (PSA) pattern, analogous to the PNA, has been recognized, albeit weaker than its North Pacific counterpart (van Loon and Shea, 1987; Karoly *et al.*, 1986). In addition, multi-decadal climate "regime shifts" are manifest in the Southern Hemisphere as in the north. The largest recent shift in the late 1970's coincided with a substantial decrease in pressure within the circumpolar trough and a concomitant delay in the seasonal breakdown of the polar vortex (Hurrell and van Loon, 1994). Chao *et al.* (2000) analyzed the pan-Pacific portion of the Global sea-Ice and SST (GISST2.2) data set and concluded that the 1976/77 Pacific basin regime shift is not unique but rather is part of a regular interdecadal oscillation that occurs about every 15 to 20 years. They suggest that north-south symmetry in the oscillation across the equator implies strong interactions between the tropics and extratropics. The available data that is used in demonstrations of symmetrical responses about the equator is highly skewed towards observations north of 10°N (Fig. 2), suggesting a need for additional climate time series from the South Pacific.

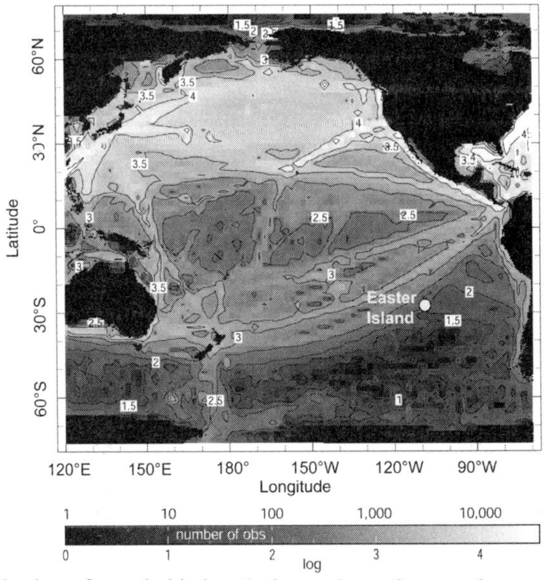

Figure 2: Number of actual ship-based observations of sea surface temperature collected between 1950 and 1975 for each 2° latitude by 2° longitude cell of the global ocean. Such observations form the core of our instrumental understanding of monthly surface ocean variability. Within most of the equatorial Pacific and the region surrounding and south of Easter Island, the black and dark gray coloration signify less than 150 measurements, e.g. from less than 50% of the total number of months between 1950 and 1975. The area directly south of Easter Island is the least observed area of the subtropical ocean, in some cases with less than 10 individual observations over this 25-year period. Corals can provide long, near-instrumental-quality reconstruction of SST from data sparse regions such as the SE Pacific.

2. EASTER ISLAND SETTING

Easter Island is located (27.1°S, 109.3°W) within the South Pacific subtropical gyre, an anticyclonic current system characterized by relatively warm stable thermal conditions, and is bathed by subtropical waters that support reef-building coral communities and incipient coral reef formations (Fig. 3). The high latitude eastward flowing West Wind Drift and the westward-flowing South Equatorial Current drive this major gyre centered between 15° to 50°S. Surface waters in its northern sector are warmed as they flow towards the west. The E-SE trades influence Easter Island for about six months of the year and N-NW winds for approximately four months. Variable to calm conditions prevail during April-May and July-August. Sea state is generally calm on the N, NW, and W shores, whereas the SE coast is often subject to high wave energy. In addition to the SE trades, the SE shore is exposed to storm-generated swell trains originating from major storm centers located between 40° to 55°S off Antarctica (Snodgrass *et al.*, 1966).

Because of the low frequency of ship traffic in the Southern Ocean, particularly in the vicinity of Easter Island, instrumental sea surface temperature (SST) data are limited (Fig. 2). The most reliable SST data available for coastal waters covers the period 1981 to present. From the composite GISST2.2 (mean = 22.8°C), NCEP/IGOSS (mean = 22.44°C), and COADS (mean = 22.76°C) data sets, representing blended ship-based and satellite observations, mean annual SST is 22.36°C (Fig. 4, Table 1). NCEP and IGOSS data obtained are identical for Easter Island therefore they will be referred to as NCEP/IGOSS throughout the paper. Between 1981 and 1999 the warmest months were February and March and the coolest months were August and September. SST values and SST anomalies (Fig. 4) reveal that during El Niño events summer temperatures at Easter Island are cooler than average, by as much as 1.7°C.

Table 1. Characteristics of monthly COADS, GISST2.2, and NCEP/IGOSS SST data, O-97-1 $\delta^{18}O$ isotope record, Easter Island air temperature, and Easter Island precipitation data between 1981 and 1995.

Mean	Minimum	Maximum	Range	Monthly statistics for 1/81- 1/95 time interval
-3.49	-2.74	-3.97	1.23	Ovahe coral $\delta^{18}O$ (^{O}/oo)
22.76	19.85	26.45	6.60	COADS SST (°C)
22.44	18.37	26.45	8.08	NCEP/IGOSS SST (°C)
22.80	18.37	25.43	7.06	GISST (°C)
20.34	16.50	24.3	7.80	Easter Air Temp (°C)
101.1	17.0	475.0	458.0	Easter Is. Precip. (mm/mo.)

Plate 1. Moai in larger than human scale.

Plate 2. Moai strewn along slopes of caldera at the quarry Rano Ravaku.

Please note that all of the color plates were taken by John T. Tanacredi and are part of a comprehensive photographic file of 887 photos in the American Museum of Natural History archives.

Plate 3. Foundation base of long houses built by Rapa Nui islanders.

Plate 4. Largest Ahu at Tongariki with several moai weighing over 90 tons.

Plate 5. Extinct volcanic caldera at Rano Raraku.

Plate 6. Rano Kau extinct volcanic caldera...primary source of drinking water on Easter Island.

Plate 7. Sooty tern islands Moto Nai and Moto Iti offshore of Easter Island observed from the rim of Rano Kau caldera.

Plate 8. Moai with coral eyes.

2.1 Coral Core Collection

In August 1997, we collected five cores of the massive coral *Porites lobata* from four locations on Easter Island. Details of the cores taken at Easter Island are summarized in Table 2. Only one of these cores exhibits distinct annual growth bands suitable for chronology development. Core Ovahe-97-1 is 1.5 m in length and was collected at 11.3 m water depth 150 m off the northeast coast (Fig. 3). There is no significant input of freshwater via runoff or rivers at Ovahe with the exception of an occasional high rainfall event during the warm season.

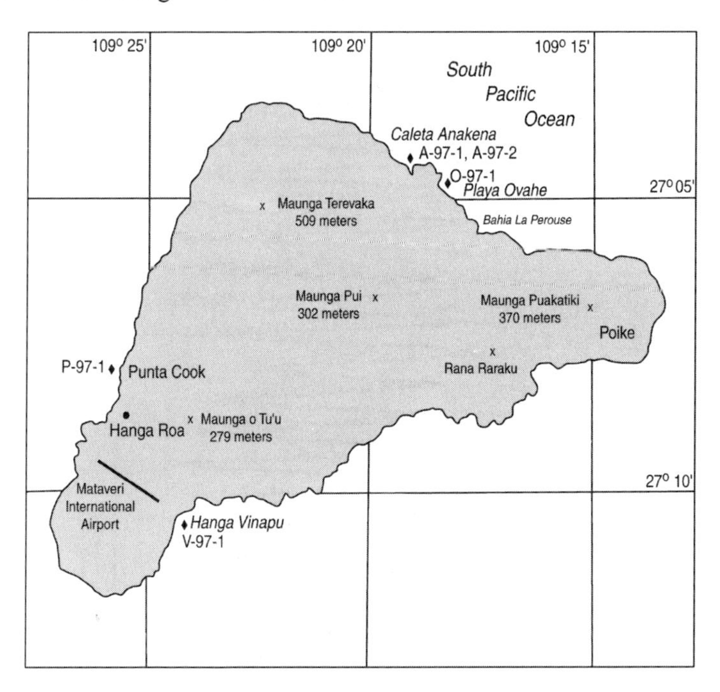

Figure 3: Easter Island is located at 27° 07'S 109° 26'W, approximately 3000 km west of Chile. Diamonds indicate the location of coral cores collected during our August 1997 expedition. We have focused on a core from Ovahe on the northern shore for our pilot studies.

Table 2: Coral cores collected from Easter Island in August 1997.

Coral Site	Core ID	Water depth (m)	Core length (m)	Species
Anakena	A-97-1	6.7	1.35	*Porites lobata*
Anakena	A-97-2	6.7	1.29	*Porites lobata*
Ovahe	O-97-1	11.3	1.50	*Porites lobata*
Punta Cook	P-97-1	18.3	2.00	*Porites lobata*
Vinapu	V-97-1	5.5	3.10	*Porites lobata*

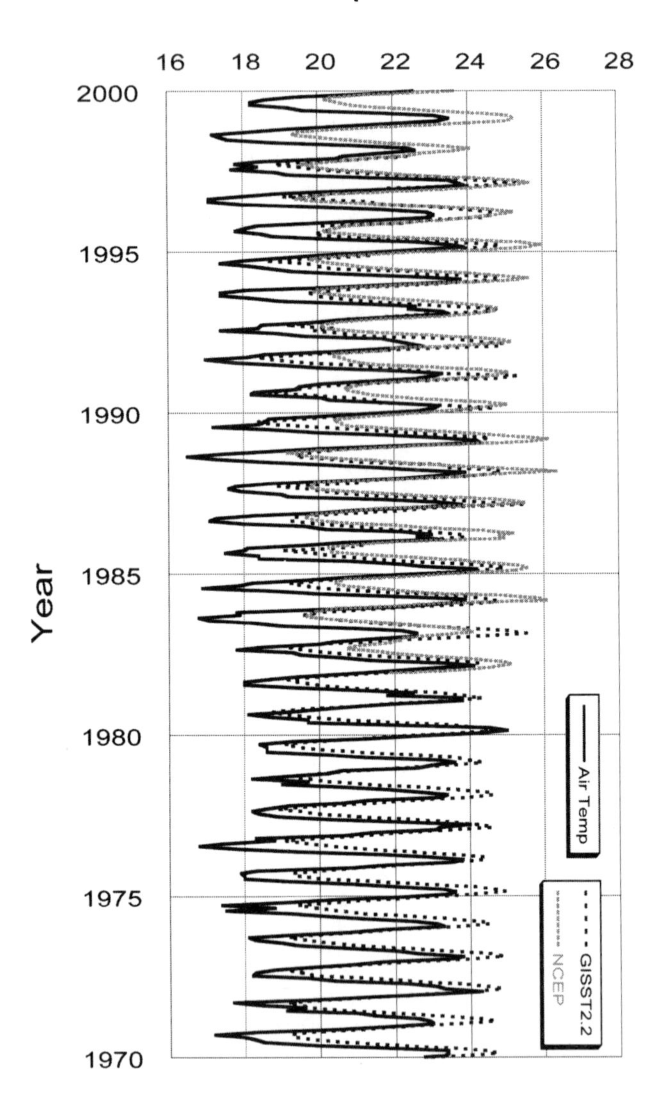

Figure 4: GISST2.2 (dashed line), NCEP/IGOSS (dotted line), and EIAT (solid line) from 1970 to 2000. Regression coefficient (r^2) for GISST2.2 vs. NCEP/IGOSS is 0.93, for GISST2.2 and NCEP/IGOSS vs. EIAT is 0.87 and 0.88, respectively. From 1944 to 2000 the regression coefficient (r^2) for GISST2.2 vs. EIAT is 0.79.

3. MATERIALS AND METHODS

All coral cores were collected using SCUBA and a hydraulic coring device equipped with a 6.5 cm diameter by 75 cm long diamond tipped core barrel. A coral core is usually made up of several sections where the lengths of individual core segments vary as a function of the quality of core recovery. After collection, all cores were curated and sun dried. Using a double-bladed brick saw, 6 mm-thick slabs were cut oriented parallel to the coral growth axis. Coral slabs were thoroughly cleaned with deionized water and air-dried. Slabs were X-rayed using Kodak Industrex AA400 film and the following parameters: 60 kV, 20 ma, and 6 s. exposure time. All of the cores except for Ovahe O-97-1exhibit highly complex growth band patterns. Subannual samples for isotopic analysis were collected from O-97-1 using a Dremel tool equipped with a 1 mm bit rotating at low speed. Coral material was collected at approximately 0.8 to 1.2 mm intervals along the growth axis forming a groove approximately 3 mm wide and 1 mm deep. Samples were stored in 0.5 ml microcentrifuge tubes for isotopic analysis. Core O-97-1 spans approximately 53 years. The average sampling density was 15 samples per year.

Aliquots of coralline aragonite weighing 55 to 95 μg were acidified in 100% phosphoric acid at 70°C for 470 s and analyzed using an automated Finnigan MAT Kiel III carbonate device coupled to a Finnigan MAT 252 isotope ratio mass spectrometer at Stanford University. A total of 868 samples were analyzed on core O-97-1 (with 8% of the samples replicated) and 155 NBS-19 (National Institute of Standards and Technology, NIST SRM 8544) standards were analyzed along with the coral samples. Replicate analysis of NBS-19 during this study yielded a standard deviation of 0.025‰ and 0.049‰ for $\delta^{13}C$ and $\delta^{18}O$, respectively. All isotope data are expressed in the conventional delta (δ) notation,

$$\delta \text{‰} = ((\text{Ratio}_{\text{sample}} - \text{Ratio}_{\text{reference}})/(\text{Ratio}_{\text{reference}})) * 1000)$$

where the isotope ratios of $^{18}O/^{16}O$ and $^{13}C/^{12}C$ are reported relative to the international VPDB (Vienna Pee Dee Belemnite) standard.

4. CALIBRATION OF OVAHE CORAL CORE 0-97-1

Coral growth bands are visible in the x-radiographs but were not clear enough throughout the entire coral core to be used as the sole means of developing an age model for the O-97-1 coral core. We therefore used a combination of banding from x-radiographs and $\delta^{18}O$ analyses interpreted in the context of seasonal variations observed in the air temperature and SST records. The GISST2.2 and NCEP/IGOSS records are monthly in resolution and are the most complete of the four SST records, albeit in part because of their use of infilling techniques (where actual data are sparse) and blending of satellite information with surface observations. The fourth SST data set (COADS) is dependent on specific measurements and has an average of less than 5 observations per year from the grid point closest to Easter Island between 1974 and 1992 and less than 6 observations from 1968 to 1992. This poses a problem for calibration of the coral record but also highlights the short and incomplete nature of actual instrumental observations in the Easter Island region. Given the limited number of COADS SST observations per year, data from this record was only used as a comparator with the GISST2.2 and NCEP/IGOSS monthly resolution data sets. Differences between monthly SST anomalies among the instrumental data sets are related to the number of observations. Root mean square differences of >0.6°C are common in parts of the Southern Hemisphere and tropical Pacific (Hurrell and Trenberth, 1999) where observations are sparse (Fig. 2). The GISST data prior to 1981 has large errors due to infilling from sparse and distant actual data. The three SST data sets span different times and all have weaknesses (Hurrell and Trenberth, 1999). The only instrument data actually collected on Easter Island is a monthly air temperature record (Easter Island air temperature (EIAT)) extending from 1941 to the present. In developing the age model maximum $\delta^{18}O$ values were assigned to the month of the SST minimum (typically August or September). Ages in between these peaks (for approximately 15 samples/year) were assigned via linear interpolation.

The raw $\delta^{18}O$ time series was then resampled at a constant 1-month time step (12/year) using the ARAND program TIMER for direct comparison with monthly instrumental records. Due to differences in record length between the COADS, GISST2.2, and NCEP/IGOSS SST, and the EIAT instrumental record, the common time period 1981 to 1995 was used for statistical comparisons. Prior to 1981, COADS, GISST2.2, and NCEP/IGOSS SST estimates for the Easter Island region exhibit large offsets, reflecting increased uncertainty in the older data. During 1981 to 1995, monthly coral $\delta^{18}O$ is highly correlated with air temperature (r^2 = 0.65), SST from GISST ($r^2 = 0.70$), and NCEP/IGOSS ($r^2 = 0.71$) as shown in Table 3. The NCEP SST record exhibits the largest seasonal variation in temperature with a total range of 8.1°C (18.37 to 26.45°C), similar to the air temperature range of 7.8°C (16.5 to 24.3°C), and the *in-situ* water temperature range of 7°C (19.5 to 26.5°C recorded at Easter Island during

1999 to 2000 (Wellington *et al*., 2001). The air temperature record is the only complete instrumental record that spans the entire $\delta^{18}O$ data set. In order to cross-calibrate our coral $\delta^{18}O$ values with instrumental air temperature and then with SST data, we first examined the relationship between EIAT and SST records from 1981 to 1995. High r^2 values for regressions of air temperature with GISST (0.87) and NCEP/IGOSS (0.88) indicate that air temperature closely tracks SST. Comparing the entire $\delta^{18}O$ time series with EIAT yields good monthly (r^2 = 0.55) and quarterly (r^2 = 0.39) correlations from 1944 to 1997. The total range in $\delta^{18}O$ from 1944 to 1997 is 1.23‰ (-2.74 ‰ to -3.97 ‰). Coral $\delta^{18}O$ regressed against NCEP/IGOSS SST from 1981-1995 (monthly) yields a slope of -0.16 ‰ per 1°C, about 25% lower than expected from theory. Monthly SSS values close to Easter Island were obtained from NCEP and Levitus *et al.* (1994). NCEP SSS exhibits a larger range (0.47‰) than the Levitus *et al.* (1994) monthly climatology (0.25‰). Both data sets show a positive correlation between SSS and SST. In the Pacific, $\Delta\delta^{18}O_{seawater}/\Delta$Salinity ranges from 0.12 to 0.27‰ per 1‰ salinity (Dunbar and Wellington, 1981; Fairbanks *et al* 1997; Linsley *et al.,* 1999). We estimate an impact on coral $\delta^{18}O$ ranging from 0.03 to 0.06‰ (using Levitus *et al*., 1994 salinities) or from 0.06 to 0.12‰ (using NCEP salinities). The potential impact of salinity-induced variability in seawater $\delta^{18}O$ is therefore less than 10% of the total $\delta^{18}O$ range in core O-97-1. Examination of the precipitation record from Easter Island shows no distinct wet season and no significant correlation when compared to the $\delta^{18}O$ record or with any of the SST or air temperature data sets. The monthly average rainfall from 1974 to 1998 is 103.2 mm/month (12.2 to 475 mm) with an annual average rainfall of 1105 mm/yr (503 to 1900 mm). The absence of any significant correlation between coral $\delta^{18}O$ and Easter Island rainfall as well as regional NCEP salinity data indicates that Easter Island coral $\delta^{18}O$ is controlled primarily by SST.

Table 3. Linear monthly regression coefficients (as r^2) between NCEP/IGOSS SST, COADS SST, GISST2.2 monthly SST, Easter Island air temperature, and Ovahe-97-1 $\delta^{18}O$ and $\delta^{13}C$ between 1981 and 1995. Shading indicates a correlation of negligible significance.

NCEP/IGOSS SST	GISST2.2	Easter Is. Air Temp	COADS SST	Monthly correlations as r^2, 1/81- 1/95
0.71	0.70	0.65	0.60	Ovahe coral $\delta^{18}O$ (‰)
0.19	0.17	0.22	0.12	Ovahe coral $\delta^{13}C$ (‰)
	0.93	0.88	0.79	NCEP/IGOSS SST (°C)
		0.87	0.78	GISST (°C)
			0.75	Easter Air Temp (°C)

5. RESULTS AND DISCUSSION

5.1 Coral Banding

The 81 cm O-97-1 *Porites lobata* core is estimated to span 53 years extending from mid 1944 to mid 1997. Because of the complex skeletal growth structure in one portion of the coral core we have the greatest confidence in the chronology from 1973 to 1997. Correlating the $\delta^{18}O$ record with the EIAT record we have extended the time series to mid 1944 with a 1.5 year hiatus at 1972 -73. Preliminary $\Delta^{14}C$ (bomb pulse detection) data for O-97-1 supports this chronology (Warren Beck, pers. comm.). Coral skeletal density (or growth) banding is clearest in the uppermost portion of the core (Fig. 5). Many species of hermatypic corals produce density bands (Knutson *et al.*, 1972). Low and high-density bands are formed by changes in the relative rates of coral skeleton extension and calcification. Environmental and biological parameters such as temperature, light, nutrient availability, turbidity, salinity, and reproductive status have all been suggested as controls on the formation of low- and high-density bands (Hudson *et al.*, 1976; Wellington and Glynn, 1983; Dunbar and Cole, 1993; Dunbar *et al.*, 1994; Wellington and Dunbar, 1995; Wellington *et al.*, 1996; Druffel, 1997). Comparison of the SST and $\delta^{18}O$ records with the density bands in O-97-1 show that the low-density bands form during the warm season beginning around December. This is different from observations in other locations such as gulfs of Chiriquí and Panamá where low density bands in corals form in the dry season when light levels are high and SST temperatures are relatively low (Wellington and Glynn, 1983; Wellington and Dunbar, 1995). At Galápagos (Dunbar *et al.*, 1994; Wellington *et al.*, 1996) and in Kenya (Dunbar and Cole, 1995; Cole *et al.*, 2000), high-density bands appear at the beginning of the annual warm season, before the highest temperatures are achieved. At Caño Island, Costa Rica, low-density bands form during the dry season when light levels are high, with SST rising (but still low) producing depleted $\delta^{18}O$ values (Wellington and Dunbar, 1995).

Depending on the species, most temperate and tropical corals between 35°N and 32°S grow optimally at temperatures between 20 and 26°C and have thermal tolerance limits ranging from a low of 16 to 18°C and a high of 30 to 32°C (Stoddart, 1969; Coles and Jokiel, 1977; Druffel 1997). SST at Easter Island ranges between 18 and 27°C (averaging 22.5°C). Given this range, high temperature thermal stress rarely, if ever, occurs whereas low temperature stress may be more common. The Ovahe coral may be exposed to optimal temperature and light conditions that promote the formation of low-density growth bands due to rapid coral extension during the warmest months of the year. During the cool season, Easter Island corals are subjected to temperatures near their low temperature limit (and likely reduced light intensity due to increased winter cloud cover and water turbidity) leading to the formation of high-density bands.

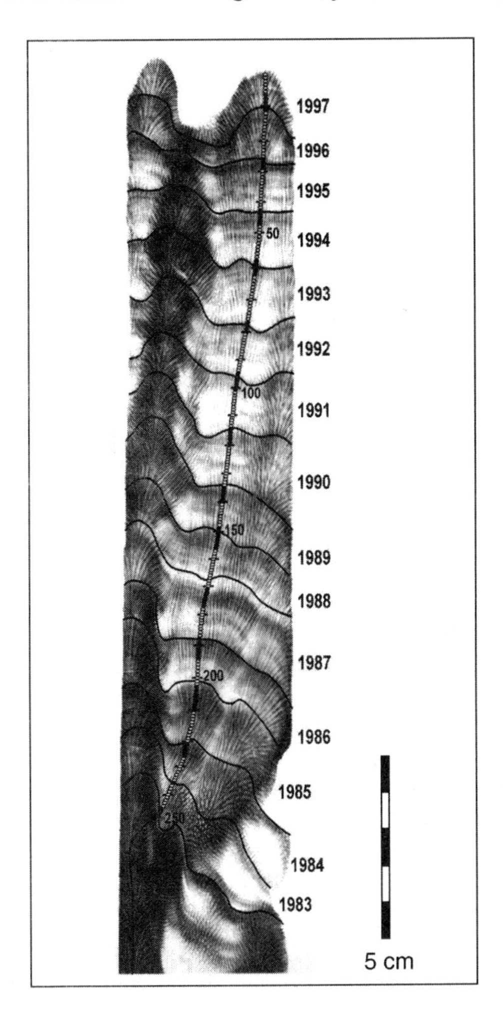

Figure 5: X-radiograph positive of uppermost part of Easter Island core O-97-1 spanning the interval from 1982 to 1997. Black and white circles represent samples taken from high and low density areas for isotopic analyses. Comparison of the SST and $\delta^{18}O$ isotope records with the density bands in O-97-1 show that the low-density bands are formed during the warm season (beginning around December). The analyzed portion of this core extends to 1944. Below this level, the growth banding is difficult to interpret.

Monthly average rainfall from 1974 to 1998 at Easter Island is 103.2 mm/month (a range of 12.2 to 475 mm). Annual average rainfall is 1105 mm (a range of 503 to 1900 mm). The precipitation record shows no distinct wet season and no correlation with the $\delta^{18}O$ record or with any of the SST or air temperature data sets. Although there is no direct monthly-scale correlation between precipitation, SST, or $\delta^{18}O$, there is a tendency for lower rainfall at Easter Island during the warm season.

5.2 Oxygen Isotope

Corals secrete calcium carbonate as the mineral aragonite during skeletogenesis. The $\delta^{18}O$ of this aragonite varies with seawater temperature and isotopic composition. In theory, carbonate $\delta^{18}O$ decreases by ~0.22‰ for every 1°C increase in water temperature (Epstein *et al.*, 1953; Grossman and Ku, 1986). Corals precipitate aragonite out of isotopic equilibrium with seawater, complicating the relationship between $\delta^{18}O$ and SST. Calibration studies have shown that the degree of isotopic disequilibrium can vary by several parts per mil but is generally constant within a coral species (Weber and Woodhead, 1972; Dunbar and Wellington, 1981, 1995; Wellington *et al.*, 1996). In areas where there are large seasonal and interannual changes in SST, $\delta^{18}O$ isotope records have successfully been used to track regional SST at sub-annual resolution (Fairbanks and Dodge, 1979; Dunbar and Wellington, 1981; McConnaughey, 1989; Wellington, *et al.*, 1996). The isotopic composition of seawater also affects the coral $\delta^{18}O$. In many areas of the tropical and subtropical oceans these variations are relatively small, but they can become significant in regions where precipitation, evaporation, and runoff influence the $\delta^{18}O$ of the seawater.

The Ovahe $\delta^{18}O$ record exhibits a seasonal cycle from 1973 to 1997 (Fig. 6) with seasonal mimima and maxima of about –3.96 and –2.75‰. There is also a distinct seasonal cycle in the SST records from the GISST2.2 and NCEP/IGOSS (Fig. 4). The NCEP/IGOSS SST record exhibits the largest seasonal variation in temperature with a total range of 8.08°C with a maximum and minimum from 18.37 to 26.45°C. Correlating the monthly GISST2.2 and NCEP/IGOSS SST records from 1981 to 1995 with the O-97-1 $\delta^{18}O$ record over the same time span yields the following equation:

$$T\,(^{\circ}C) = -6.1792\,(\delta^{18}O\ ‰_{\text{aragonite}}) + 0.9085 \qquad r^2 = 0.70$$

This simple linear regression yields a slope equivalent to -0.16‰ in $\delta^{18}O$ per 1°C, a value lower than the 0.22‰ per 1°C expected from theory. Although this slope value lies within the full range of slopes reported for

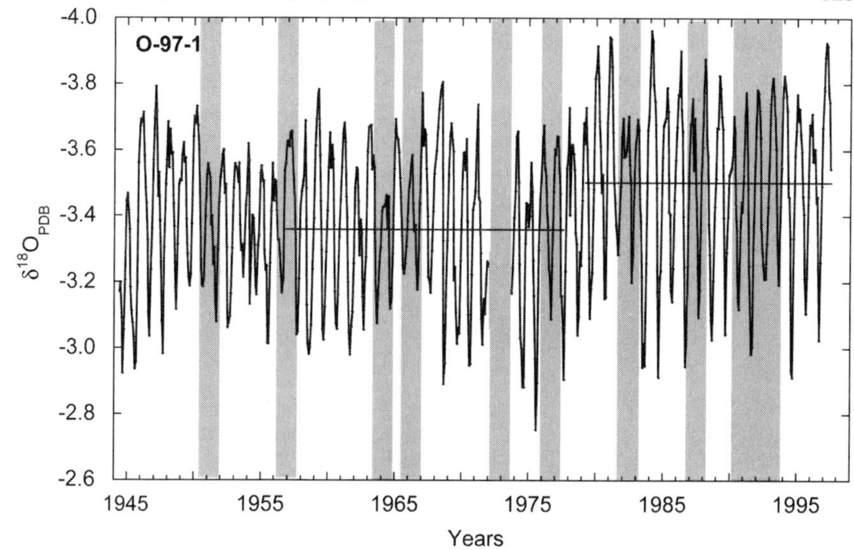

Figure 6: Monthly-resolution (~15 samples/year) $\delta^{18}O$ stratigraphy of coral core O-97-1 from Ovahe, Easter Island. Shaded regions indicate ENSO warm events of the past 55 years. Note that each ENSO warm mode year is characterized by a reduced amplitude of the annual cycle of SST, consistent with the observation that ENSO warm events produce cooler summers but warmer winter and spring conditions in this part of the Pacific. This expression of ENSO means that sub-annual sampling is necessary as the signal is obscured in annual average data sets.

Porites from many sites across the Pacific (Weber and Woodhead, 1972; Patzöld, 1984; McConnaughey, 1989; Cole *et al.*, 1993; Linsley *et al.*, 1994; Tudhope *et al.*, 1995; Linsley *et al.*, 1999; Wellington and Dunbar, 1995; Wellington *et al.*, 1996; Quinn *et al.*, 1998; Evans *et al.*, 2000), the possibility remains that there is some influence by seasonal variability in seawater $\delta^{18}O$ on our calibration. Unfortunately, we have no instrumental salinity or seawater $\delta^{18}O$ data from Easter Island so it is difficult to directly assess the impact of seawater isotopic variability on the coral record. However, the NCEP and Levitus *et al.* (1994) salinity data in addition to the Easter Island precipitation provide some insights. Monthly average rainfall from 1974 to 1998 is 103.2 mm/month (a range of 12.2 to 475 mm). Annual average rainfall is 1105 mm (a range of 503 to 1900 mm). The precipitation record shows no distinct wet season and no significant correlation with the monthly $\delta^{18}O$ record or with any of the SST or air temperature data sets. Nevertheless, there is a tendency for lower amounts of rain to fall at Easter Island during the warm season. Increased temperature will yield lower coral $\delta^{18}O$ values while increased salinity will elevate coral $\delta^{18}O$. Even with a $<10\%$ contribution of $\delta^{18}O_{seawater}$ to the coral isotope signal as a function of the small range in SSS, it is possible that the coral $\delta^{18}O$ signal is partially obscured by the counteracting effects of temperature and salinity, consistent with our observations of a $\delta^{18}O$ response slope less than that predicted by theory. Since the $\delta^{18}O$ isotope record from core O-97-1 has a large seasonal variation that correlates well with the SST records since 1981, it is possible that the isotope record is a better indicator of variations in SST at Easter Island than the sparse instrumental data sets prior to 1981. In fact, we find a stronger relationship between $\delta^{18}O$ and EIAT (r^2 of 0.52) than between GISST2.2 and EIAT ($r^2 = 0.41$) or between GISST2.2 and $\delta^{18}O$ ($r^2 = 0.45$) from 1944 to 1995.

Two important features can be observed in the Ovahe coral $\delta^{18}O$ record. The 1976 "climate shift", known mainly from North Pacific indices (Ebbesmeyer *et al.*, 1991; Deser *et al.*, 1996; Mantua *et al.*, 1997) is apparent in the Ovahe record as a post-1976 depletion in warm season $\delta^{18}O$ values of about 0.10 to 0.15 ‰ relative to the preceding 30 years. This result suggests Southern Hemisphere participation in this event, consistent with the observations of Chao *et al.* (2000). The extent to which previous climate regime shifts occur in the southeastern Pacific and are correlative with Northern Hemisphere events can be determined using long coral records from Easter Island. In addition, ENSO "warm-mode" years are expressed as cooler summers and warmer winters and springs in the Easter coral record (Fig. 6 shows this reduced annual cycle), consistent with the modern observational pattern (Fig. 7). A multi-taper spectral analysis (Fig. 8) of this short record reveals decadal and ENSO-band peaks, with most variance concentrated at the annual cycle and quasi-biennial oscillation (QBO) periods.

5.3 Climate Implications

The ENSO system dominates interannual variability of the ocean and atmosphere in the tropical and subtropical Pacific. This signal propagates through the global atmosphere leaving its imprint on planetary systems as diverse as Antarctic sea ice, maize yields in Africa, and rainfall in Florida. Although theories of ENSO physics have recently provided a basis for prediction (Latif *et al.*, 1998; Neelin *et al.*, 1998), this understanding comes primarily from observations collected since the late 1970's. Instrumental and paleoceanographic data as well as modeling studies suggest that the post-1970's ENSO system is unusual (Trenberth and Hoar, 1996; Hughen *et al.*, 1999; Federov and Philander, 2000), so the extent to which ENSO is predictable over longer time scales is not yet known. The ENSO warm

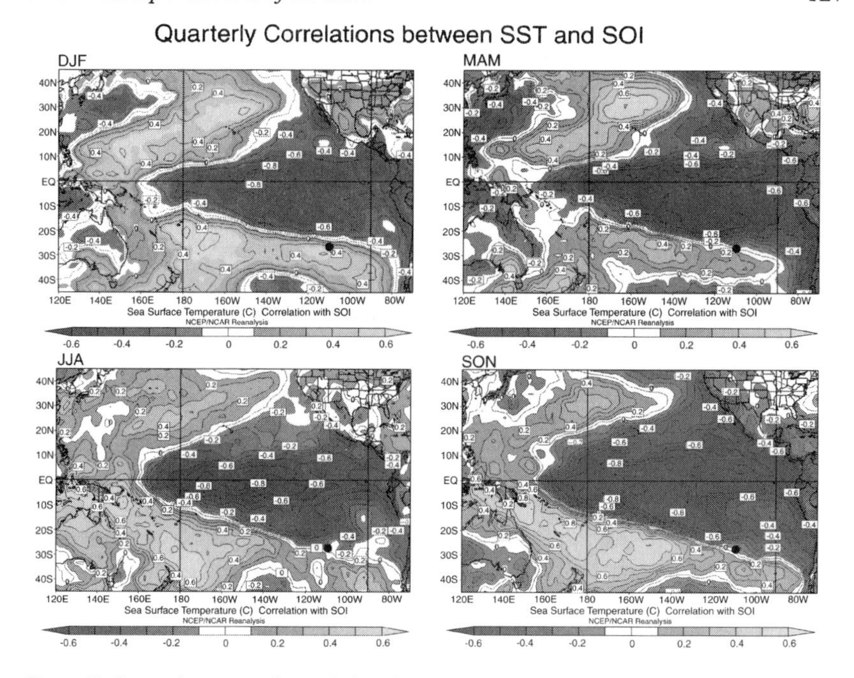

Figure 7: Quarterly averaged correlations between SST and SOI for the Pacific Ocean using the NCEP/NCAR reanalysis SST products and spanning the period 1950-1995. The location of Easter Island is shown on each quarterly panel. Note that during ENSO warm events Easter Island lies in a region of cool SST's during months DJF and warm SST's during months JJA and SON.

events of 1982/83 and 1997/98 were the strongest of the past century (McPhaden, 1999). Their magnitudes were not correctly predicted nor was the extended moderate ENSO episode of 1990-1995 (Trenberth and Hoar, 1996). In fact, using current models, it is apparent that there are substantial differences in ENSO forecast skill from decade to decade (Balmaseda *et al.*, 1995; Chen *et al.*, 1995,). The cause of these changes is not clear but leading candidates include a change in the base state of the tropical climate system and a change in the phase of the decadal ENSO-like climate anomaly (e.g., a 1976/77 style regime shift). Based on theoretical considerations, Federov and Philander (2000) suggest that recent anomalous ENSO events may reflect a change in the background or time-averaged climate state and include the possibility that the system is responding to global warming. The interdecadal variability described in the Pacific by Chao *et al.* (2000) includes four "steps" during the 20th century. These steps are also accompanied by changes in ENSO. Our Easter Island coral record supports Chao *et al.* (2000) analysis by confirming the presence of a 1976/77 climate step at Easter Island. Longer coral records from Easter can be used to study earlier "steps", during periods when instrumental data is unavailable. One outstanding question is the extent to which anomalous ENSO's of the 1980's and 1990's are unique and related to global warming versus a "normal" expression of decadal variability. Additional long coral records from key areas of the Pacific will help answer these questions. Easter Island is well suited for tracking both ENSO and the mean state of the eastern South Pacific Gyre. Because ENSO events are manifest at Easter Island as reductions in the seasonal cycle as opposed to large annual excursions, we can track ENSO-related interannual variability without obscuring small changes in mean state.

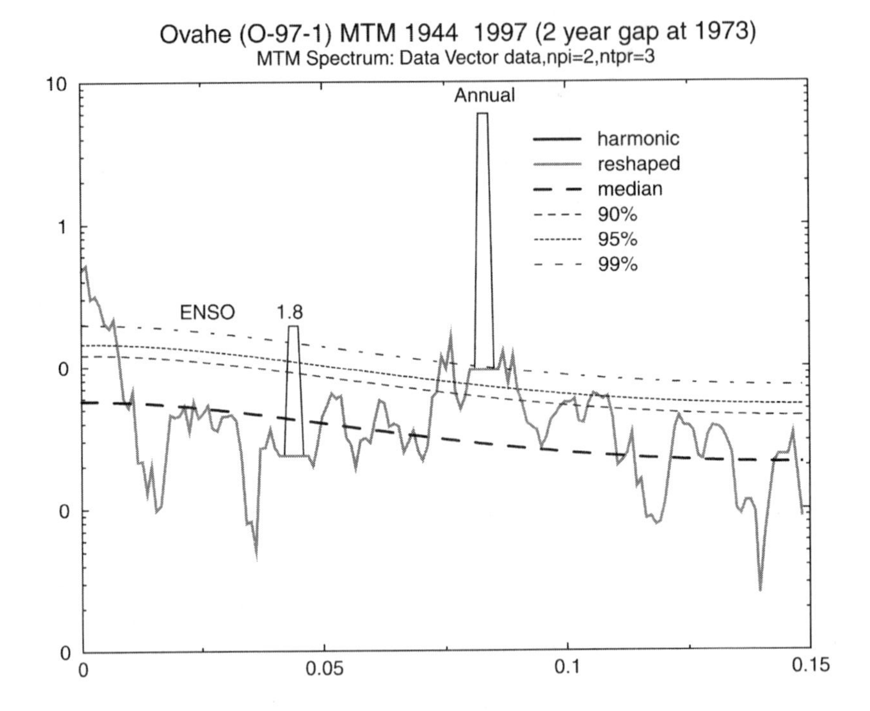

Figure 8: Spectral plot of O-97-1 $\delta^{18}O$ record from 1944 to 1997. The spectral results show a significant peak at one year, indicative of a seasonal signal. The peak at 1.8 years suggests a quasi-biennial oscillation (QBO) signal. There is a less pronounced peak present at 3.9 years that represents variability in the ENSO band.

6. CONCLUSIONS SUMMARIZED

The recognition of large-scale coherent climate patterns such as ENSO and the NAO has led to increased predictive skill by regional and global climate models. Most large-scale systems described thus far are in the Northern Hemisphere, largely because of the greater availability and length of historical climate data sets. Coherent climate patterns occur in the Southern Hemisphere as well - the newly described Southern Ocean Wave is one likely candidate. However, in much of the Southern Hemisphere and particularly in the Pacific Ocean, the short and sparse instrumental data limit the analysis of climate systems operating at time scales greater than interannual. The Ovahe O-97-1 coral isotope record contributes to knowledge of Southern Hemisphere variability by providing a coral-based time series of sea surface temperatures from Easter Island. Located at 27.1°S, 109°W, Easter Island lies within the southeastern Pacific subtropical gyre approximately 3000 km from Chile. Few instrumental records are available from this region and these tend to be short, spanning no more than the last 26 years. Easter Island lies within the descending limb of the Pacific Walker cell, a location where sea surface temperatures influence the Southern Oscillation. El Niño events result in an attenuated annual cycle in SST, a signal that is captured by coral oxygen isotope values. Decadal variability is also pronounced. The O-97-1 coral record shows that a 1976/77 Pacific basin climate shift, well-known from the tropics and the North Pacific, is represented as a significant warming step at Easter Island, consistent with theories that tropical forcing contributes to decadal variability within the extratropics. The $\delta^{18}O$ calibration against the various SST, SSS, and the instrumental EIAT data suggests that the O-97-1 oxygen isotope record from Easter Island is a reliable indicator of past variability in SST. We estimate that coral records from Easter Island have the potential to span at least 250 years, and possibly 400 years, greatly extending the instrumental data set available for analysis of interannual-to-decadal climate variability in the South Pacific.

ACKNOWLEDGMENTS

We especially thank John Loret at the Museum of Long Island for organizing the expedition to Easter Island and all of the Explorer's Club participants for their interest and support. Additional thanks to Warren Beck, George Burr, Michel Garcia, and William Kempner for their drilling and field assistance. We thank Mike Lutz for his assistance in the laboratory. We greatly appreciate the government of Chile for granting permission to conduct this research at Easter Island. Financial support for this project was partially provided by the Explorer's Club.

REFERENCES

Chao, Y., M. Ghil, J.C., McWilliams. 2000. Pacific Interdecadal Variability in This Century's Sea Surface Temperatures, *Geophys. Res. Lett.*, 27, 2261-2264.

Chen, D., S.E., Zebiak, A.J., Busalacchi, and M.A., Cane, 1995, An improved procedure for El Niño forecasting: Implications for predictability, *Science*, 269, 1699-1702.

Cole, J.E., R.G. Fairbanks, and G.T. Shen, 1993, Recent variability in the Southern Oscillation: Isotopic results from a Tarawa Atoll coral, *Science*, 260, 1790-1793.

Cole, J.E., R.B. Dunbar, T.R. McClanahan, and N.A. Muthiga, 2000, Tropical Pacific forcing of decadal SST variability in the western Indian Ocean over the past two centuries, *Science*, 287, 5453, 617-619.

Coles, S.L., and P.L. Jokiel. 1977, Effects of temperature on photosynthesis and respiration in hermatypic corals. *Mar. Biol.*, 43, 209-216.

Dreser, C., M.A. Alexander, and M.S. Timlin. 1996, Upper-ocean thermal variations in the North Pacific during 1970-1991, *J. Climate*, 9, 1840.

Druffel, E.R.M, 1997, Geochemistry of corals: Proxies of past ocean chemistry, ocean circulation, and climate, *Natl. Acad. Sci.*, 94, 8345-8361.

Dunbar, R.B., and J.E. Cole, 1993, Coral records of ocean-atmosphere variability, NOAA Climate and Global Change Program Special Report #10, 38 p.

Dunbar, R. B., and J. E. Cole, 1995, The tropical influence on Global Climate: Are surprises the rule? EOS, Transactions, *Am. Geopys. Union*, 76, 43, p. 431.

Dunbar, R.B., and G.M. Wellington, 1981, Stable Isotopes in a branching coral monitor seasonal temperature variations, *Nature*, 293, 453-455.

Dunbar, R.B., G.M. Wellington, M.W. Colgan, and P.W. Glynn, 1994, Eastern Pacific sea surface temperature since 1600 A.D.: The $\delta^{18}O$ record of climate variability in Galapagos corals. *Paleoceanography*, 9, 2, 291-315.

Ebbesmeyer, C.C., D.R. Cayan, D.R. McClain, F.H. Nichols, D.H. Peterson, and K.T. Redmond, 1991, 1976 step in Pacific climate: Forty environmental changes between 1968-1975 and 1977-1984, In: Proc. 7th Annual Pacific Climate (PACLIM) Workshop, April, 1990, California Department of Water Resources, J.L. Betancourt and V.L. Tharp (editors), Interagency Ecological Study Program Technical Report 26, 115-126.

Epstein, S., R. Buchsbaum, H.A. Lowenstam, and H.C. Urey, 1953, Revised carbonate-water isotopic temperature scale, *Bull. Geol. Soc. Am.*, 64, 1315-1326.

Evans, M.N., A. Kaplan, and M.A. Cane, 2000, Intercomparison of coral oxygen isotope data and historical sea surface temperature (SST): Potential for coral-based SST field reconstruction, *Paleoceanography*, 15, 5, 551-563.

Fairbanks, R.G., and Dodge, R.E., 1979, Annual periodicity of the $^{18}O/^{16}O$ and $^{13}C/^{12}C$ ratios in the coral *Montastrea annularis*, *Geochim. Cosmochim, Acta*, 43, 1009-1020.

Fairbanks, R.G., M.N. Evans, J.L. Rubenstone, R.A. Mortlock, K. Broad, M.D. Moore, C.D. Charles, 1997, Evaluating climate indices and their geochemical proxies measured in corals, *Coral Reefs,* 16, suppl., 93-100.

Federov, A.V., and S.G. Philander, 2000, Is El Niño changing?, *Science*, 288, 1997-2002.

Grossman, E.L., and T.L. Ku, 1986, Oxygen and carbon isotope fractionation in biogenic aragonite, *Chem. Geol.*, 59, 59-74.

Gu, D., and S.G.H. Philander, 1997, Interdecadal climate fluctuations that depend on exchanges between the tropics and extratropics, *Science*, 275, 805-807.

Hudson J.H., E.A. Shinn, R.B. Halley, and B. Lidz, 1976, Sclerochronology - a tool for interpreting past environments, *Geology*, 4, 360-364.

Hughen, K.A., J.T. Overbeck, L.C. Peterson, S. Trumbore, 1999, Rapid climate changes in the tropical Atlantic region during the last deglaciation, *Nature*, 380, 51-54.

Hurrell, J.W., and H. van Loon, 1994: A modulation of the atmospheric annual cycle in the Southern Hemisphere, *Tellus*, 46A, 325-338.

Hurrell, J.W., and K.E. Trenberth, 1999: Global Sea Surface Temperature Analyses: Multiple Problems and Their Implications for Climate Analysis, Modeling and Reanalysis, *Bull. Amer. Meteor. Soc.*, 80, 2661-2678.

Ji, M., A. Leetmaa and J. Derber, 1995: An ocean analysis system for seasonal to interannual climate studies, *Mon. Wea. Rev.*, 123, 460-481.

Karoly, D.J., G.A. Kelley, J.F. Le Marshall, and D.J. Pike, 1986, An atmospheric climatology of the southern hemisphere based on ten years of daily numerical analyses (1972-1982). WMO Long-Range Forecasting Research Report No. 7, WMO/TD, No. 92.

Knutson, D.W., R.W. Buddemeier, and S.V. Smith, 1972, Coral chronologies: seasonal growth bands in reef corals, *Science*, 177, 270-272.

Latif, M., D. Anderson, T. Barnett, M. Cane, R. Kleeman, A. Leetmaa, J. O'Brien, A. Rosati, E. Schneider, 1998, A review of the predictability and prediction of ENSO, *J. Geophys. Res.*, 103, 14,375-14,394.

Levitus, S., 1982, Climatological Atlas of the World Ocean. NOAA Prof. Paper No. 13 U.S. Dept of Commerce, Washington, D C.,173 p.

Levitus, S., R. Burgett, T. P. Boyer, 1994, World Ocean Atlas 1994, Vol. 3, Salinity, NOAA Atlas NESDIS, 3, Dept. of Commerce, Washington, D. C., 99 p.

Linsley, B.K., R.B. Dunbar, G.M. Wellington, and D.A. Mucciarone, 1994, A coral-based reconstruction of Intertropical Convergence Zone variability over Central America, *J. Geophy. Res.*, 99, C5, 9977-9994.

Linsley, B.K., R.G. Messier, R.B. Dunbar, 1999, Assessing between-colony oxygen isotope variability in the coral Porites lobata at Clipperton Atoll, *Coral Reefs*, 18, 13-27.

Mantua, N.J., S.R. Hare, Y. Zhang, J.M. Wallace and R.C. Francis, 1997, A Pacific interdecadal climate oscillation with impacts on salmon production, *Bull. Amer. Meteorl. Soc.*78, 1069-1079.

McConnaughey, T.A., 1989, 13C and 18O isotopic disequilibrium in biological carbonates: I, Patterns, *Geochim. Cosmochim. Acta*, 53, 151-162.

McPhaden, M.J., 1999, Genesis and evolution of the 1997-98 El Niño, Science, 283, 950-954.

Neelin, D.J., D.S Battisti, A.C Hirst, F.F. Jin, Y., Wakata, T.Yamagata, and S.E. Zebiak, 1998, ENSO theory, *J. Geophys.Res.*, 103, 14261-14290.

Patzöld, J., 1984, Growth rhythms recorded in stable isotopes and density bands in the reef coral *Porites lobata* (Ceba, Philippines), *Coral Reefs*, 3, 87-90.

Quinn, T.M., T.M. Crowley, F.W. Taylor, C. Henin, P. Joannot, and Y. Join, 1998, A multicentury stable isotope record from a New Caledonia coral: Interannual and decadal sea surface temperature variability in the southwest Pacific since 1657 A.D., *Paleoceanography*, 13, 412-426.

Snodgrass, F.E., G.W. Groves, K.F. Hasselman, G.R. Miller, W.H. Munk, and W.H. Powers, 1966, Propagation of ocean swell across the Pacific, *Phil. Trans. Roy. Soc.*, London, 259, 431-497.

Stoddart, D.R., 1969, Ecology and morphology of recent coral reefs, *Biol. Rev.* 44, 433-498.

Trenberth, K., and T. Hoar, 1996, The 1990-1995 El Niño-Southern Oscillation event: Longest on record, *Geophys. Res. Lett.*, 23, 57-60.

Tudhope, A.W., G.B. Shimmield, C.P. Chilcott, m. Jebb, A.E. Fallick, and A.N. Dalgleish, 1995, Recent changes in climate in the far western equatorial Pacific and their relationship to the Southern Oscillation: oxygen isotope records from massive corals, *Earth Planet. Sci. Lett.*, 136, 575-590.

Van Loon, H. and D.J. Shea, 1987: The Southern Oscillation. Part VI: Anomalies of sea level pressure on the Southern Hemisphere and of Pacific sea surface temperature during the development of a warm event, *Mon. Wea. Rev.,* 115, 370-379.

Weber, J.N., and P.M.J. Woodhead, 1972, Temperature dependence of oxygen-18 concentrations in reef coral carbonates, *J. Geophy. Res.*, 77, 463-473.

Wellington, G.M., and P.W. Glynn, 1983, Environmental influences on skeletal banding in eastern Pacific (Panamá) corals, *Coral Reefs*, 1, 215-222.

Wellington, G.M., and R.B. Dunbar 1995, Stable Isotopic signature of El Nino-Southern Oscillation events in eastern tropical Pacific reef corals, *Coral Reefs*, 14, 5-25.

Wellington G.M., R.B. Dunbar, and G. Merlen, 1996, Calibration of stable isotope signatures in Galapagos corals, *Paleoceanography*, 11, 4, 467-480.

White, W.B., and R.G. Peterson, 1996, An Antarctic circumpolar wave in surface pressure, wind, temperature, and sea ice extent, *Nature*, 380, 699-702.

Zhang, R.H., L.M. Rothstein, A.J. Busalacchi, 1998, Origin of upper-ocean warming and El Niño change on decadal scales in the tropical Pacific Ocean, *Nature*, 391, 879-883.

Chapter 8

Prehistoric Destruction of the Primeval Soils and Vegetation of Rapa Nui (Isla de Pascua, Easter Island)

D. MANN[1], J. CHASE[2], J. EDWARDS[3], W. BECK[4], R. REANIER[5], and M. MASS[3]

[1]Institute of Arctic Biology, University of Alaska, Fairbanks, AK 99775; [2]New York University, School of Medicine, New York, NY 10016; [3]Oregon Health & Science University, Portland, OR 97239; [4]AMS Facility, Department of Physics, University of Arizona, Tucson, AZ 85721; [5]Reanier and Associates, 1807 32nd Ave, Seattle, WA 98122

1. INTRODUCTION

Traditional knowledge and the earliest archaeological ^{14}C date suggest that people arrived on Rapa Nui as early as A.D. 400 (Smith, 1961; Heyerdahl and Ferndon, 1961; Ayers, 1971; Bahn and Flenley, 1992). Prior to human arrival, much of the island was probably forested, with the largest trees a now-extinct species of palm (Flenley *et al.*, 1991). Between A.D. 1000 and 1700, the Rapa Nui people erected megalithic sculpture, may have used a written language, and possibly numbered >10,000 people (Bahn, 1993). However, when James Cook visited Rapa Nui in 1774, he found only several thousand people eking out a living amidst ruins on an island barren of trees. In a hypothesis originated by Mulloy (1970) and fully developed by Flenley and King (1984), Flenley *et al.* (1991), and Bahn and Flenley (1992), uncontrolled population growth destroyed the natural vegetation, degraded the island's ecosystems, and eventually led to the near extinction of the human population. This putative ecological history of Easter Island is cited as support for predictive models for the human use of natural resources (Brander and Taylor, 1998) and has passed into the modern folklore embodied by popular cinema. Accepting that Rapa Nui is a microcosm for

Easter Island, Edited by John Loret and John T. Tanacredi
Kluwer Academic/Plenum Publishers, New York, 2003

the planet Earth, the "Lost Eden" interpretation of its history paints a grim picture of our collective future (Bahn and Flenley, 1992). On the other hand, the paleoecological data that describes the ecological history of Rapa Nui is far from complete (Orliac and Orliac, 1998; Nunn, 2000). As detailed below, the lake-sedimentary records described so far from the island are only roughly dated, and they contain suggestions of major stratigraphic unconformities. A persistent question in Rapa Nui prehistory has been when the destruction of the primeval forest began. As pointed out by La Perouse (1799), the loss of a sheltering forest on a windy oceanic island initiates drastic ecological changes that can be reversed only with great difficulty.

Our purpose here is to document the timing of earliest vegetation clearance by humans on Rapa Nui and its impacts on the island's soils. Soils are key for ecosystem function and their loss can have catastrophic consequences. Soils are the medium of plant growth; consequently, their fertility and hydrologic characteristics strongly influence primary productivity and vegetation type. They pass through marked developmental stages spanning millennia, during which they equilibrate with climate, topography, and biota into quasi-steady states. When these equilibria are upset, for instance by climate change or human disturbance, re-equilibration may take centuries to millennia.

2. STUDY AREA

Today the interior of Rapa Nui is a largely treeless landscape dominated by bunch grasses growing in fields of shattered lava that surround the rounded slopes of volcanic cones (Fig. 1). Plantations of exotic eucalyptus trees thrive at all altitudes on the island, but coconut palms grow poorly at most sites. The modern landscape is crisscrossed by stock fences and stonewalls, which date to sheep-herding times.

There are only two lakes, Rano Raraku and Rano Kau, both of which are located in volcanic craters. A third crater, Rano Aroi, contains a reed swamp. The island's one permanent surface stream originates from Rano Aroi and flows through a partly unroofed cave for a kilometer before disappearing underground. The island is riddled with caves, and brackish springs are common along the coastline (Heyerdahl, 1961). Dry stream courses descend the slopes of Terevaka in several places, becoming less developed as they approach the sea. Some contain plunge pools, and debris-flow levees occur along some channel reaches. In general, alluvial fans are poorly developed. Our overall impression, based on features of fluvial geomorphology, is that surface runoff on Rapa Nui has been sporadic and generally slight during the Quaternary.

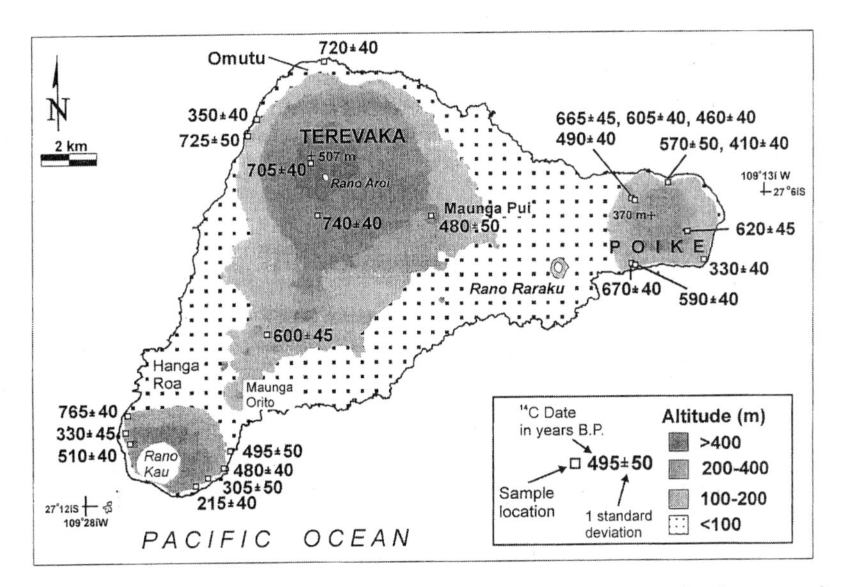

Figure 1. Rapa Nui (Isla de Pascua, Easter Island) showing the three major volcanic centers of the island and the locations and ages of ^{14}C dated charcoal collected in this study. ^{14}C ages are uncalibrated and are shown with one sigma errors.

2.1 Location and Climate

Rapa Nui (Fig. 1) lies in the southeastern Pacific Ocean at 27° 8' S latitude, 109° 26' W longitude. Central Florida is located at a similar latitude in the northern hemisphere. Rapa Nui is a small island – its surface area is only 120 km^2. It is volcanic in origin, and its coastline is rocky and cliffed. Cool waters prevent the formation of a fringing coral reef. Mean annual temperature is 21°C, with January the warmest month at 24° C and August the coolest month at 18° C (International Station Meteorological Climate Summary, 1995). Rainfall averages 1130 mm *per annum* with a minimum in August and September (Streten and Zillman, 1984); however, rainfall is quite variable on an annual basis and its historical variations are poorly documented. The island is swept continuously by wind, which, combined with the cool temperature and the absence of a sheltering forest discourages the growth of tropical plants.

Rapa Nui is a marginal environment for traditional Polynesian dryland farming (Stevenson *et al.*, 1999). Banana trees require special wind shelters there. In prehistoric times, a large number of sweet potato (*Ipomoea batatas*) varieties were raised on Rapa Nui and formed the staple food (Yen, 1974; Cummings, 1998). Other aboriginal cultivars included banana, sugar cane, *Colocasia* and *Alocasia* taro, *Dioscorea* yams, gourds, *Triumfetta semitroba* (a source of cordage), and the paper mulberry (Hyerdahl, 1961). In many fields, lithic mulches were used to conserve soil moisture and to provide shelter from wind and sun (La Perouse, 1799; Wozniak, 1998). La Perouse noted that the Rapa Nui people added the ashes of burned vegetation to augment soil fertility. Indication of the surface aridity of Rapa Nui is given by the fact that plans for agricultural improvement in modern times stipulate irrigation using water pumped from the crater lakes and from deep wells (Porteous, 1981).

2.2 Vegetation

The modern vegetation of Rapa Nui consists of about 200 species, most of which were introduced (Flenley *et al.*, 1991). Only eight plant species survive that are endemic to the island, and native species number only fortysix (Skottsberg, 1956). Trees are rare and all are introduced species. Introduced grass species dominate the vegetation below about 400m, while native sedges and bunch grasses are widespread above this altitude (Flenley *et al.*, 1991). Today, people use fire to maintain the lowland grasslands for grazing. Except for the plantations of eucalyptus trees, the grassy landscape of Rapa Nui that we see today probably is similar to what the earliest European visitors saw in the A.D. 1700s (Orliac and Orliac, 1998). Although Polynesian farmers disrupted the major primeval vegetation associations on the island, numerous plant species probably survived on cliffs and in ravines only to be extirpated by sheep in the A.D. 1800s.

2.3 Soils

A variety of different types and ages of soils occur on Rapa Nui, all of which have been disturbed by prehistoric agriculture and historic sheep grazing. Sizable (>50 ha), contiguous areas of agriculturally productive soils occur today on the western slopes of Poike, around the town of Hangaroa, near Vaihu on the southeast coast, on the plain southwest of Rano Raraku, and at Vaitea in the island's interior (Wright and Diaz, 1962). Small pockets of arable soils occur scattered across the rest of the island. As described below, intensive erosion has stripped the primeval soil from areas where the soil cover was formerly thickest.

Rapa Nui's soils fall into three general categories according to parent material and age. On the peninsula of Poike, old lava flows have weathered deeply, producing oxisols (ferrallitic soils) enriched in iron oxides and neoformed clays. Prolonged weathering has reduced underlying basalt flows to saprolite to depths of >5 m in some areas of Poike. The high clay content of these soils and the high porosities created by the root casts of the native palms (see below) endow them with high water-retention capacities. When the surface of these oxisols are denuded of vegetation and dried, extensive wind erosion occurs, as is happening now in blowouts on the southeastern side of Poike. Though the oxisol profiles on Rapa Nui are now truncated, similar soils in other subtropical regions possess A horizons containing acid mull that is the source for most of the soil's cation-exchange capacity (Soil Conservation Service, 1975; Mikhailov, 1999). When this surface horizon is lost through erosion, the soil's fertility declines markedly (Duchafour, 1977).

The second widespread soil type consists of hydrothermally altered deposits of clay and oxidized scoria on the flanks of recently active volcanic vents. Mikhailov (1999) suggests that all of the island's red soils result from the hydrothermal alteration of the volcanic bedrock. We think this underestimates the amount of pedogenesis that has occurred on the island's older volcanic landforms, but we agree with Mikhailov (1999) about the importance of hydrothermal processes in generating red clays along the line of youthful scoria cones stretching between Rano Kau and Terevaka.

The third widespread soil type on Rapa Nui includes the drought-prone Inceptisols and Entisols developing on the stony surfaces of lava flows that issued from Terevaka within the last several thousand years old (Baker, 1967). Slope wash and aeolian transport has concentrated fine material in topographic low points within these flows. In these small basins and on footslopes, Entisols are forming on progressively accumulating deposits of loess and slope wash. Inceptisols show development of a mollic epipedon while entisols are too disturbed by sediment input to show any horizonation. Archaeological surveys (e.g., Stevenson *et al.*, 1999) document widespread use of these soils for agriculture, and we can infer that most were modified by tillage.

2.4 The Crater Lakes

Volcanoes form the three corners of Rapa Nui. Their lava flows are primarily basalt, hawaiite, mugearite, and benmoreite, with minor amounts of rhyolite (Bandy, 1937; Baker, 1997). The oldest rocks on the island are the 0.5 myr basalt/hawaiite flows (Hasse *et al.*, 1997) forming the Poike stratovolcano at the eastern corner of the island. Poike's small summit crater, Puakatike, today contains a eucalyptus grove. In 1998, we used a soil auger to probe in this crater to a depth of 4 m but found only well-drained silty clay. Poike has been severely eroded by the sea on all four sides, and was once an island that was later connected to Terevaka by lava flows issuing from the latter (Baker, 1967).

The southwestern corner of the island is formed by another volcano, which contains the caldera lake of Rano Kau. The Rano Kau volcano was active until ca. 0.3 myr (Hasse *et al.*, 1997). This volcano is intermediate between Poike and Terevaka in the degree of its sea-cliff and soil development.

Terevaka forms most of the landmass of Rapa Nui and supports numerous parasitic cones on its flanks. Lava flows and scoria cones arranged along volcanic fissures link Terevaka to the two older volcanoes. The broad, gently sloping dome of its summit reaches 500 m altitude. Terevaka has a complex eruptive history with most of its mass built by a series of lava flows and pyroclastic eruptions from near its present summit (Baker, 1967).

The small cone of Raraku is important for the present discussion because of the sediment cores taken from its crater lake, Rano Raraku. The lake is enclosed in a crater of composite origin. The crater's southeastern side is composed of tuff from which the moai were carved. A spectacular cliff was eroded in this tuff by the sea before lava flows from Terevaka shifted the coastline several kilometers further to the northeast. The Raraku tuffs dip northwestward, indicating that the former vent lay to the southeast. The western rim of the crater is largely scoria and postdates the tuff eruption.

Rano Raraku is slightly elongate to the north and measures 300 m from east to west. Several transects with an echo sounder revealed that the lake is flat-bottomed with water depths of 6-7 m in its central basin. Totora reeds (*Scirpus californicus*) covered approximately 1/6 of the lake surface in 1998. Some of these reeds are rooted in a band ringing the lakeshore, while the larger portion float on the lake surface and are redistributed from side to side by the wind. Rano Raraku has been frequently photographed, and these photographs indicate that over the last century there has never been a complete reed mat covering the lake, and that the locations of the floating mats are continuously changing. In this way, the vegetation mats in Rano Raraku are different from the stable mats familiar to us around infilling ponds at higher latitudes.

2.5 Paleoecology

The present-day fauna of Rapa Nui is impoverished. Bones recovered from excavations at Ahu Naunau reveal that at least six species of land birds were indigenous to Easter Island when the Polynesians arrived, including at least one species of rail (Steadman, *et al.*, 1994; Steadman, 1995). Several species of sea birds that no longer frequent the island occur in archaeological deposits. Several species of introduced birds, together with horses, cattle, cats, dogs, and the black rat range across the island today. Sheep, cattle, and pigs were introduced by missionaries ca. 1866. Though rare today, approximately 40,000 sheep were grazing the island in 1943 (Porteous, 1981).

Most of the primeval flora of Rapa Nui was extirpated through a combination of Polynesian agriculture and European sheep grazing. Evidence from palynology (Flenley *et al.*, 1991) and charcoal analysis (Orliac, 2000) indicates the presence of an unexpectedly diverse woody flora that was still present on the island between the 14th and 17th centuries AD. Orliac (2000) speculates that a closed, mesic forest similar to that occurring today on islands in eastern Polynesia probably existed on Rapa Nui prior to Polynesian arrival. Several of the now extinct or locally extirpated tree species like the extinct palm (*Paschalococos disperta*) and *Aliphitonia zizyphoides* probably grew to large size. Their presence suggests that the primeval forest, at least locally, was more than low-statured scrub (Flenley, 1993).

Pioneering work on Easter Island's ecological history was done by John Fleney (Flenley and King, 1984; Flenley *et al.*, 1991; Flenley, 1993). Fleney retrieved sediment cores from the island's three crater lakes and analyzed them for pollen, described their chemical stratigraphy, and obtained radiocarbon dates. Though invaluable as a comprehensive reconnaissance study, Flenley's data do not provide close dating control on possible human-associated ecological events. Similarly, the most recent contribution to the paleo-ecology of Rapa Nui (Dumont *et al.*, 1998) fails to establish a firm chronological base for ecological changes.

A close look at Flenley's data show that sediments dating to the time of Polynesian settlement are either missing or highly disturbed in the cores he analyzed from the crater lakes. In his core from Rano Raraku, sediment samples that were ^{14}C-dated to 480 and 6850 years B.P. are only 15 cm apart in the stratigraphy (Flenley, 1993). A recent investigation of Rano Raraku's sediments by Dumont *et al.* (1998) also suggest a major unconformity in the stratigraphy immediately prior to 590 ± 60 ^{14}C yr B.P. (Gif-9629). In Rano Aroi, a ^{14}C age of 19,000 years was obtained near the core top, while a modern ^{14}C age came from a depth of 1 m (Flenley, 1993). In Rano Kau, ages of 1000 yr B.P. were obtained on sediments 5 m apart in the core. There are no continuous records of ecological change over the last 2000 years on Easter Island.

Flenley's data do tell us several interesting things about the ecological history of Easter Island. First, in prehistoric times the island was largely forested, in stark contrast to the present day. Second, in the undisturbed, pre-2000 year B.P. portions of the most detailed pollen diagram (Rano Aroi), large fluctuations occurred in the percentages of palm, Compositae, grasses, and ferns during the Holocene. These vegetation changes must have been caused by climatic change, since humans were not yet present on Rapa Nui.

3. METHODS

Stratigraphic sections were prepared by clearing slumped material to expose undisturbed sediments. Soil profiles were described in naturally occurring erosion gulleys and in soil pits. Plant macrofossils and charcoal fragments were extracted from stratigraphic sections for AMS-radiocarbon dating by washing material through 500- and 150-micron sieves and examining the residues under a dissecting microscope.

Lake cores were taken with a gravity corer about 110 m from the southern end of Rano Raraku in 6 m of water. We worked off the edge of a floating totora mat. Sediment compaction shortened both cores to roughly half of their undisturbed lengths. Age control in the lake cores comes from AMS-^{14}C dates on the seeds of totora (*Scirpus californicus*). We separated these seeds from sediments by washing sediments through a 250-micron sieve and then sorting the plant debris under a dissecting microscope. We identified totora seeds from ancient sediments by comparing them to those of living totora plants from Rapa Nui and coastal California. Organic content was determined using loss on ignition after an hour of heating at 550° C (Dean, 1974). Charcoal abundance was determined using the sieving method of Long *et al.* (1998). ^{14}C dates were calibrated using Calib4 (Stuiver *et al.*, 1998).

4. RESULTS

4.1 The Primeval Soils

Brightly colored oxisols containing palm root casts are the primeval soils on land surfaces located outside the areas covered by recent lava flows and scoria cones (Fig. 2). Palm roots leave distinctive casts in clayey soils because they are parallel sided and unbranched (Flenley, 1993) (Fig. 3). On Rapa Nui these root casts are several millimeters to a centimeter in diameter, commonly contain traces of carbonized material, and can occupy up to 40-50% of soil volume. The high clay content and the creation of high porosities by palm root casts endow Rapa Nui's primeval soils with high water-holding capacity. On the southern slopes of Terevaka, palm root casts are absent above about 300 m altitude. On the upper slopes of Terevaka, the primeval soil consisted of a 30-80 cm thick A horizon in which finely divided, black organic matter is mixed with fine-grained tephra.

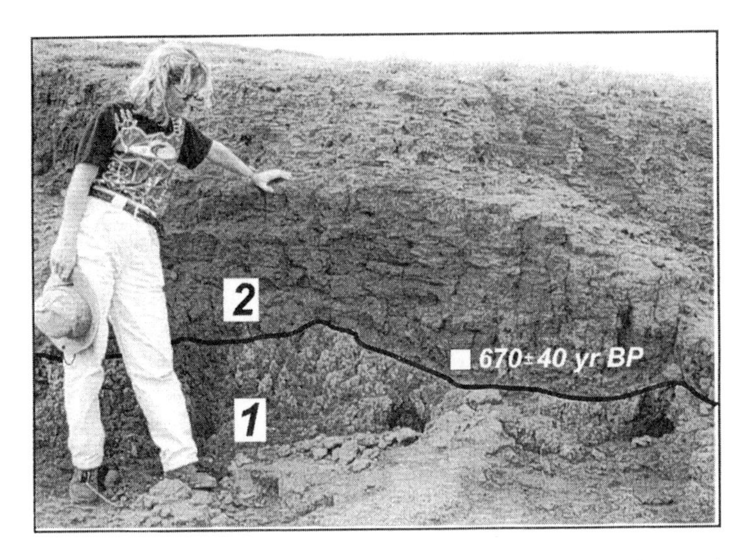

Figure 2. Section "16Aug98b2" on the southwest side of Poike. Unit 1 is the primeval oxisolic soil containing abundant palm root casts. White box is the location of ^{14}C-dated, charred fragments of a palm nut near the erosional unconformity separating the primeval soil from the overlying slopewash deposits of Unit 2. ^{14}C ages are uncalibrated and are shown with one sigma errors.

Figure 3. Detail of clay-rich oxisol containing vertically oriented casts of palm roots from a barrow pit on the northern slopes of Maunga Orito.

The primeval soil is harder to identify in parts of the island covered by younger lava flows because the extant soils are often only centimeters to several decimeters thick and highly disturbed. Along the northern and western coastline of Terevaka, palm root casts do appear in fine-grained diamictons of debris-flow origin.

4.2 Soil Stratigraphy

Exposures of the primeval soil on Rapa Nui, especially in sections exposed in foot-slope areas, often reveal a prominent erosional unconformity overlain by slopewash deposits. Type sections are located within the erosion gullies forming along the inland margin of the large blowouts on the southwestern corner of Poike (27° 7.249' S, 109° 15.656 W) (Figs 1,4). Four distinctive stratigraphic units occur in these soils developed in deep (>2 m thick) unconsolidated sediments on slopes ranging from 5-40°.

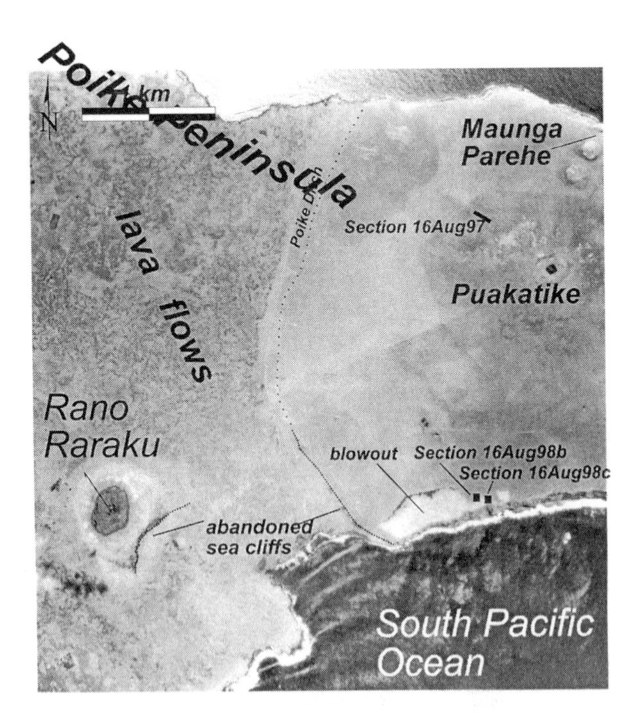

Figure 4. Vertical aerial photograph (1:25,000, 1981, Geoservice Chile, photograph number 174732) showing Rano Raraku and the western portion of Poike. Several sampling locations are shown on Poike. Note the contrast between the smooth, highly weathered surface of the Poike stratovolcano and the rough, lava flows west of the Poike Ditch.

UNIT 1. At the base of the stratigraphy are oxic horizons truncated by subsequent erosion. These oxic horizons are clay- and iron-rich, usually bright orange or red in color, and contain abundant palm root casts in their upper 2 m. They grade downwards into saprolite, rock that decayed in place through chemical weathering. Often the truncated upper surface of the primeval soil is eroded by steep-walled rills, which are up to 1 m deep and are active today.

UNIT 2. Unit 2 is a buried plaggen horizon, a soil layer modified by agricultural use. Unit 2 forms the fill material for the rills eroded downward into Unit 1 and often buries the primeval soil to depths of 10-60 cm (Fig. 4). It consists of clay pellets, pebbles, charcoal fragments, and clay rip-up clasts within a clay loam matrix. Sedimentary structures include multiple cut and fill structures suggestive of infilled erosion rills and subhorizontal pebble lines, often comprised of clay pellets, resulting from episodes of slope wash. Obsidian flakes and basalt clasts as large as small boulders occur in this unit. Charcoal fragments range up to several centimeters in size and include fragments of the nuts of the extinct palm, *Paschalococos disperta*.

UNIT 3 Overlying Unit 2 along an indistinct boundary, this unit differs from Unit 2 by lacking pebbles. Unit 3 is 40-110 cm thick and is composed of granular conglomerates of clay that are mainly in the size range of medium sand. It contains convex-up lens of silty clay demarcated by basal layers of granules. These lenses are up to 20 cm wide and 1 cm deep and probably represent episodic slope wash, but under different vegetation and/or precipitation regime than what occurred during the deposition of Unit 2. The charcoal in this unit is very fine, usually <1mm in size.

UNIT 4. The topmost unit is comprised of slightly finer-grained material than Unit 3. Charcoal is scarce but does occur; it typically is <1 mm in diameter and probably represents charred grass stems. The presence of subhorizontal lines of granules and small pebbles (clay conglomerates) suggest a slope-wash origin for Unit 4. Often black in color, Unit 4 supports the living vegetation, usually grass, and at many sites it comprises a mollic epipedon.

The slope deposit facies just described from Poike also occur along the tops of the eroding sea cliffs around Rano Kau (Fig. 5) and at the bases of slopes near Hangaroa where prolonged weathering and/or deep deposits of scoria have created great thicknesses of slope wash. Soil stratigraphy differs elsewhere on the island where the regolith is rockier and/or is underlain at shallow depths by bedrock. Along the steep northern and western flanks of Terevaka, the primeval soil (Unit 1) is sometimes evident from palm root casts, but sand and pebbles are the predominant particle sizes rather than clay. The overlying slope deposits are coarser, consisting of boulders rather than the pebbles and granules found on Poike and on the distal slopes of the

Rano Kau crater. Often only one unit is distinguishable between Units 1 and 4. Repetitive debris flow activity is evidenced at some locations on Terevaka (Fig. 6). For instance, at location "23Aug98a", pebbly, fine-grained sheet flow sediments separate two stony debris flow deposits. Organic-stained, sheet flow sediments overlying the upper stony deposit contain fragments of palm nuts. Overlying this alluvial-fan facies of the primeval soil is 20-25 cm of pebbly clay and silt, which probably represent younger, distal sheet-flow deposits.

In places the soil stratigraphy is directly associated with ancient architecture. Bluff erosion has cross-sectioned an *ahu* on the northeast flank of Maunga Parehe on the northern side of Poike (Fig. 7). The *ahu*'s foundation is composed of locally obtained stones and gravel and overlies clay-rich subsoil containing palm root casts. Mantling the *ahu* is a layer of finer-grained slope- wash material containing cut and fill structures, pebble lines, and convex-up lenses of charcoal. All these features indicate deposition by slope wash occurring after *ahu* construction.

Figure 5. Section "19Aug98b" on the western distal slopes of Rano Kau. Unit 1 with its characteristic palm root casts is truncated by an erosional unconformity. The overlying Unite 2 contains basalt boulders and slope-wash sediments of silt and clay. Arrows delineate layers of charcoal fragments. The white rectangle marks the location of unidentified wood fragments dating to 330 ± 45 yr B.P.

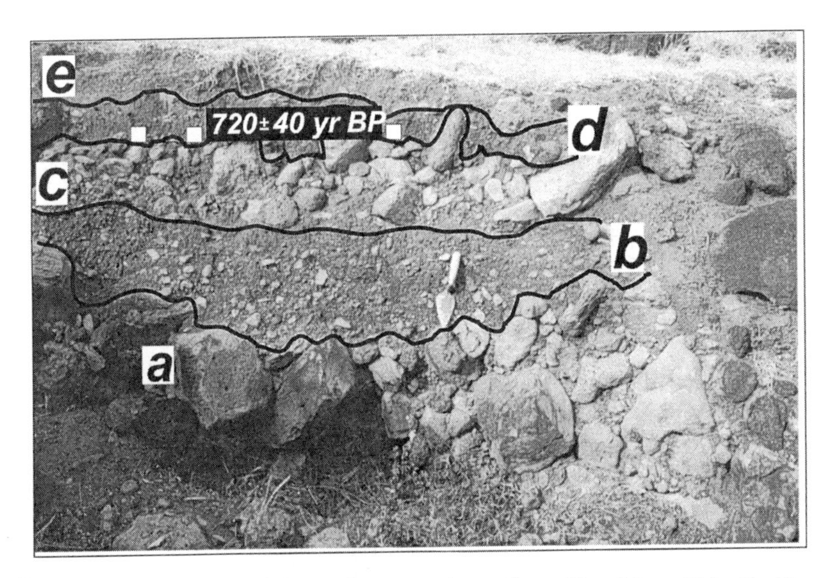

Figure 6. Section "23Aug98a" near Omutu on the northern side of Rapa Nui. Bouldery debris-flow sediments (Unit a) are overlain by finer sheet flow deposits (Unit b) and then by more debris-flow deposits (Unit c). Unit d is comprised of sheet-flow sediments containing charred fragments of palm nuts that dated to 720 ± 40 yr B.P. Unit e is composed mainly of silty clay and probably was deposited by recent slope wash. The pre-700 yr B.P. changes in depositional regime may be the result of climatic changes.

Figure 7. Photo-mosaic of the seaward, northern side of an unnamed *ahu* located on Poike near Maunga Parehe. Unit 1 is the primeval soil containing abundant palm root casts. It is intruded by the foundation of the *ahu* built out of locally quarried stones including occasional obsidian flakes. Units 3 and 4 are draped across the archaeological deposits and consist of slopewash deposits with charcoal lenses. They largely bury the upslope portions of this *ahu*.

4.3 Timing of Prehistoric Land Clearing and Soil Erosion

In the course of describing stratigraphic sections exposed by erosion, road construction, and in soil pits, we collected charcoal fragments from the slopewash sediments directly overlying the truncated primeval soil. After calibration to calendar years, these soil-charcoal dates range from A.D. 1200 to modern (Fig. 8). Surprisingly, none predate A.D. 1200. Over much of the island, the primeval soil was truncated by erosion between A.D. 1200 and 1650. From the natural exposure through the *ahu* at Maunga Parehe, we infer that widespread slope erosion occurred there after A.D. 1300-1440 (Fig. 7).

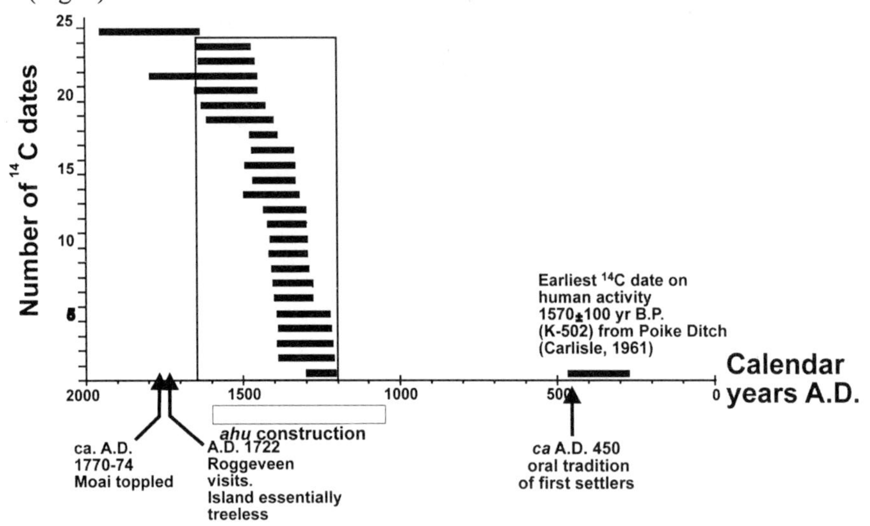

Figure 8. Comparisons between soil-charcoal dates (this study) and human history on Rapa Nui. The horizontal black bars are ^{14}C dates shown as the ranges of the two-sigma errors of their calibrated ages. Polynesian and European histories summarized by Hyerdahl and Ferndon (1961) and Bahn and Flenley (1992). Timing of *ahu* construction from Love (1993), Skjoldsvold (1993), and Stevenson et al. (1999).

4.4 A New Lake-Sediment Record from Rano Raraku

In August 1998 we raised a 2-m long core from Rano Raraku (Fig. 9). Our plan was to refine the chronology of sedimentary changes in this shallow lake (Flenley *et al.*, 1991) in order to more precisely date the timing of forest clearance and soil erosion within this crater. Starting at the 16-cm level, the core shows a striking increase in magnetic susceptibility, which is associated with the in-wash of mineral material from the surrounding slopes

(Fig. 10). The gradual rise in magnetic susceptibility above the 40 cm level corresponds to a gradual decrease in organic matter content at the same levels. Where magnetic susceptibility increases sharply at 16 cm, organic content drops. This striking change occurred at or shortly before A.D. 1070-1280 based on the AMS date from the 13-14 cm depth. Charcoal is absent or rare in the sediments below the 20-cm level. A minor peak in charcoal at the 100 cm level suggests that fires did occur on Rapa Nui before human settlement; however, the lack of charcoal over the next several millennia indicates that fire was a rare event there. Roughly coincident with the drop in organic content and the spike in magnetic susceptibility, a massive influx of charcoal occurred in Rano Raraku.

Figure 9. Rano Raraku looking north-northwest in August 1998 showing the coring site. A former, higher shoreline is evidenced by dessicated lake muds about a meter above present lake level.

The history of totora reed (*Scirpus californicus*) on Rapa Nui has been a topic of debate. Hyerdahl (1961) speculated that totora was carried to Rapa Nui by voyagers coming from South America. Flenley *et al.* (1991) found abundant Cyperaceae pollen at levels dating to >30,000 ^{14}C yr BP and suggested that a major portion of this pollen was coming from totora reeds growing on the island long before human settlement. The Cyperaceae is a large group, and the pollen is not diagnostic of individual species. More definitive evidence that totora reached Rapa Nui on its own long before human colonization comes from our AMS date of 4380 +- 50 yr BP obtained on the seeds of this species.

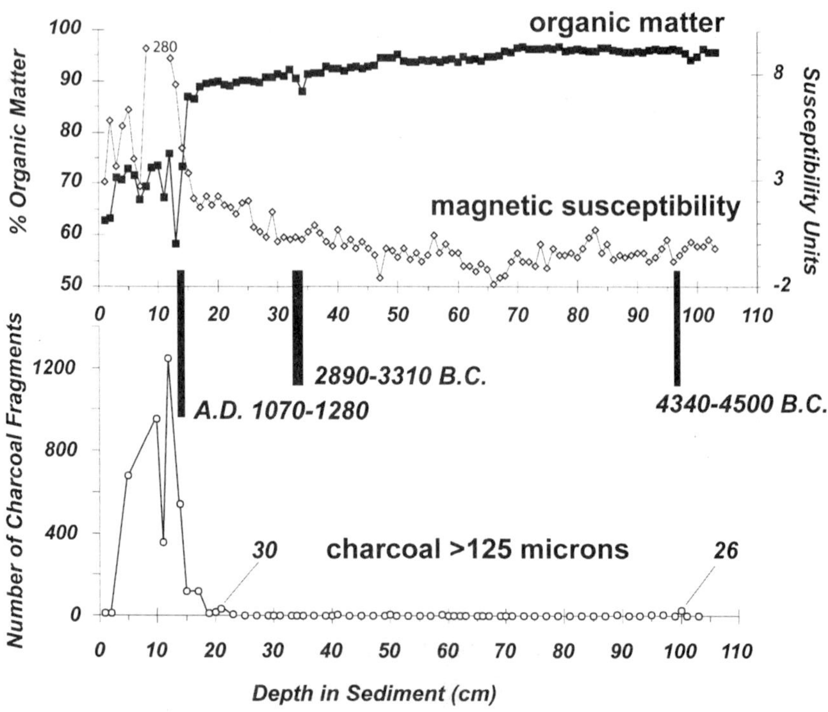

Figure 10. Stratigraphy of the Rano Raraku core. A sharp decline in organic matter content and an increase in magnetic susceptibility indicate a drastic change in the input of mineral sediments at the 16-cm level of the core. Charcoal increases at the same level. Charcoal fragments between 16 and 18 cm depth could have been mixed downward into the sediment by biological activity (root channels and borrowing animals). Note the near absence of charcoal in sediments deeper than 20 cm. Ages are from ^{14}C-dated seeds of *Scirpus californicus* and are presented as the 2-standard deviation limits of calibrated ages.

5. DISCUSSION

The health of an ecosystem is tightly linked to the health of its soils. Our results show that starting abruptly at A.D. 1200, the primeval soils of Rapa Nui were severely eroded. Abundant charcoal fragments lie at and above the erosional conformities that truncate soils containing the root casts of the extinct Rapa Nui palm. On Poike, the association between abundant charcoal, occasional flakes of obsidian originating from the other end of the island, and thick slope wash deposits is consistent with the original vegetation having been cleared by slash and burn techniques, which triggered deep and extensive slope erosion. A similar association between forest clearance by Polynesian farmers and widespread soil erosion is documented on other Pacific islands (e.g., Ellison, 1994; Kirch, 1996;

Wilmshurst, 1997; Dodson and Intoh, 1999). The presence of three distinctive slope wash units on Rapa Nui may evidence changing Polynesian agricultural practices and/or distinctive impacts on slope erosion by European sheep ranching.

The similarity of oldest charcoal dates from different parts of Rapa Nui (Fig. 8) suggests that initial forest clearance and soil erosion occurred everywhere on the island within several centuries and certainly within the 450-year interval between A.D. 1200 and 1650. Chronological precision is limited by the errors inherent in the radiometric analysis and by the fact that the age of the burned plants pre-date the fire by several years, or possibly, in the case of trees, by several centuries.

There are no clear geographic trends in the timing of the initiation of soil erosion on Rapa Nui (Fig. 1). The age of charcoal buried under slope-wash sediments high on Terevaka is similar to that found on Poike and on the distal slopes of Rano Kau. Charcoal abundance in our lake-sediment core from Rano Raraku increases sharply at the level corresponding to A.D. 1070-1280, agreeing with the timing indicated by soil-charcoal dates elsewhere on the island. The sharp increase in mineral material in Rano Raraku sediments after A.D. 1000 reflects the same radical increase in erosion that is indicated by the soil stratigraphy exposed in erosion gullies scattered across the island.

Widespread burning and soil erosion began on Rapa Nui only after A.D. 1200, yet archaeological evidence suggests that initial Polynesian settlement occurred at least 400 years earlier (Stevenson, 1995) and probably 700-900 years earlier (Stevenson *et al.*, 1999; Kirch, 2000). It is striking that our dates suggest that the island remained unburned and presumably uncleared of vegetation until as late as A.D. 1200, and the charcoal record from Rano Raraku suggests that fires were rare or absent prior to ca. A.D. 1200.

Perhaps little land was cleared of native vegetation prior to A.D. 1200 simply because human populations were small until then. If little land were being cleared, then burning, slope erosion, and charcoal deposition would also have been of limited extent. This seems unlikely because Rapa Nui has a high potential for wildfire spread. Its vegetation probably was never a perennially lush, evergreen tropical forest where fires had difficulty spreading. Furthermore, its topography is simple, it is intermittently droughty, and it is subject to strong winds. These characteristics make Rapa Nui an ideal setting for extensive wildfires, even if it were covered in mesic forest (Orliac, 2000). If small areas of the island were being cleared by slash and burn techniques, occasional fires would have escaped control and left records in the soil stratigraphy or in the sediments of Rano Raraku. We interpret the soil-stratigraphic and lake-sedimentary records described above as indicating that forest clearance began on Rapa Nui only after A.D. 1200.

Why was widespread forest clearance delayed for centuries after initial settlement? Island colonization histories often may involve shifts between an early phase characterized by opportunistic and locally contingent subsistence strategies that is followed by more specialized, higher productivity modes (Yen, 1990). We speculate that the Polynesian occupation of Rapa Nui passed through two distinct phases. Between ca. A.D. 300 and A.D. 1200, the island was occupied, perhaps transiently, by people making a living by hunting and gathering along its rocky shores and by exploiting its flightless bird populations (cf., Steadman, 1997). The second phase started only ca. A.D. 1200 when dry land farming was initiated in the island's interior and large human population could be supported. It remains unknown whether this switch in economy was triggered by climatic changes (c.f., Nunn, 2000) or by cultural innovations, which possibly involved the introduction of new, dryland food crops like sweet potatoes (Yen, 1974; Kirch, 2000). Additional paleoecological data is required to test this and all previous speculations concerning the history of this particular island microcosm.

ACKNOWLEDGEMENTS

We thank John Loret, Dorothy Peteet, the Science Museum of Long Island, and the Explorer's Club for organizing and funding the field trips to Rapa Nui during which these investigations were conducted. Sergio Rapu and John Tanacredi greatly assisted with field logistics. Joan Wozniak provided useful discussions in the course of this project. Bruce Finney and Andrea Krumhardt provided lake coring equipment and technical expertise. We thank Jim Smyth and Bill Kempner for exemplary assistance in the field.

REFERENCES

Ayres, W.S. (1971). Radiocarbon dates from Easter Island. *Journal of the Polynesian Society* 80, 497-504.

Bahn, P.G. (1993). The history of human settlement on Rapanui. pp. 53-55 In: "Easter Island Studies" (S.R.Fischer, ED.). Oxbox Monograph 32, The Short Run Press, Oxford, England.

Bahn, P.G. and Flenley, J.R. (1992). "Easter Island, Earth Island." Thames and Hudson, London.

Baker, P.E. (1967). Preliminary account of recent geological investigations on Easter Island. *Geological Magazine* 104, 116-122.

Baker, P.E. (1998). Petrological factors influencing the utilization of stone on Easter Island. pp. 279-283, In: "Easter Island in Pacific Context South Seas Symposium. Proceedings of the Fourth International Conference on Easter Island and East Polynesia". University of

New Mexico, Albuquerque, 5-10 August 1997. C.M. Stevenson, G. Lee, and F.J. Morin, Eds. The Easter Island Foundation, Bearsville and Cloud Mountain Press, Los Osos, CA.

Bandy, M.C.(1937). Geology and petrology of Easter Island. *Geological Society of America Bulletin* 48, 1589-1610.

Brander, J.A. and Taylor, M.S. (1998). The simple economics of Easter Island: A Ricardo-Malthus model of renewable resource use. The American Economic Review 88, 119-138.

Cummings, L.S. (1998). A review of recent pollen and phytolith studies from various contexts on Easter Island. pp. 100-106, In: "Easter Island in Pacific Context South Seas Symposium. Proceedings of the Fourth International Conference on Easter Island and East Polynesia". University of New Mexico, Albuquerque, 5-10 August 1997. C.M. Stevenson, G. Lee, and F.J. Morin, Eds. The Easter Island Foundation, Bearsville and Cloud Mountain Press, Los Osos, CA.

Dean, W.E. (1974). Determination of carbonate and organic matter in calcareous sediments and sedimentary rocks by loss on ignition: comparison with other methods. *Journal of Sedimentary Petrology* 44, 242-248.

Dodson, J.R. and Intoh, M. (1999). Prehistory and paleoecology of Yap, federated states on Micronesia. Quaternary International 59, 17-26.

Dransfield, J., Flenley, J.R., King, S.M., Harkness, D.D., and Rapu, S. (1984). A recently extinct palm from Easter Island. *Nature* 312, 750-752.

Duchafour, P. (1977). "Pedology: Pedogenesis and Classification". George Allen and Unwin, London.

Dumont, H.J., Cocquyt, Fontugne, M., Arnold, M., Reyss, J-L., Bloemendal, J., Oldfield, F., Steenbergen, C.L.M., Korthals, H.J., and Zeeb, B.A. (1998). The end of moai quarrying and its effect on Rano Raraku, Easter Island. *Journal of Paleoliminology* 20, 409-422.

Ellison, J.C. (1994). Paleo-lake and swamp stratigraphic records of Holocene vegetation and sea-level changes, Mangaia, Cook Islands. *Pacific Science* 48, 1-15.

Flenley, J.R. (1993). The present flora of Easter Island and its origins. Pp. 7-15, In: "Easter Island Studies" (S.R. Fischer, ED.). Oxbox Monograph 32, The Short Run Press, Oxford, England.

Flenley, J.R. (1993). The paleoecology of Easter Island, and its ecological disaster. pp. 27-45, In: "Easter Island Studies" (S.R. Fischer, ED.). Oxbox Monograph 32, The Short Run Press, Oxford, England.

Flenley, J.R., King, S.M., Jackson, J., and Chew, C. (1991). The Late Quaternary vegetational and climatic history of Easter Island. *Journal of Quaternary Science* 6, 85-115.

Flenley, J.R. and King, S.M. (1984). Late Quaternary pollen records from Easter Island. *Nature* 307, 47-50.

Hasse, K.M., Stoffers, P., and Garbe-Schonberg, C.D., 1997. The petrogenetic evolution of lavas from Easter Island and neighboring seamounts, near-ridge hotspot volcanoes in the southeast Pacific. *Journal of Petrology* 38, 785-813.

Heyerdahl, T. and Ferdon, E.N. (eds). (1961). "Reports of the Norwegian Archaeological Expedition to Easter Island and the East Pacific", Vol. 1, "Archaeology of Easter Island". Forum Publishing House, Stockholm, 559 pp.

International Station Meteorological Climate Summary. (1995). Volume 3.0. Table 42, "Foreign Station Climatic Summary". United States Federal Climate Complex, Department of Commerce, Asheville, North Carolina, (compact disc).

Kirch, P.V. (1996). Late Holocene human-induced modifications to a central Polynesian island ecosystem. Proceedings of the National Academy of Sciences 93, 5296-5300.

Kirch, P.V. (1997). Microcosmic histories: Island perspectives on "Global" change. American Anthropologist 99, 30-42.

Kirch, P.V. (2000). On the Road of the Winds: An Archaeological History of the Pacific Islands before European Contact." University of California Press, Berkely.

La Perouse, J. F.(1995). "The Journal of Jean-Francois de Galaup de La Perouse, 1785-1788". Hakluyt Society; London.

Long, C.J. , Whitlock, C.Bartlein, P.J., and Millsaugh, S.H., 1998. A 9000-year fire history from the Oregon Coast Range, based on a high-resolution charcoal study. *Canadian Journal of Forest Research* 28, 774-787.

Nunn, P.D. (2000). Environmental catastrophe in the Pacific Islands around A.D. 1300. *Geoarchaeology* 15, 715-740.

Orliac, C. and Orliac, M. (1998). The disappearance of Easter Island's forest: over-exploitation or climatic catastrophe? pp. 129-134, In: : "Easter Island in Pacific Context South Seas Symposium. Proceedings of the Fourth International Conference on Easter Island and East Polynesia". University of New Mexico, Albuquerque, 5-10 August 1997. C.M. Stevenson, G. Lee, and F.J. Morin, Eds. The Easter Island Foundation, Bearsville and Cloud Mountain Press, Los Osos, CA.

Orliac, C. (2000). The woody vegetation of Easter Island between the eartly 14th and the mid-17th centuries AD. Pages 211-220 In: C.M. Stevenson and W.S. Ayres (Eds.), "Easter Island Archaeology: Research on Early Rapa Nui Culture." Easter Island Foundation.

Porteous, J.D. (1981). "The Modernization of Easter Island". Western Geographical Series Volume 19, Department of Geography, Universty of Victoria, Victoria, British Columbia, Canada. 304 pp.

Skottsberg, C., 1956. "The Natural History of Juan Fernandez and Easter Island". Volume 1. Almqvist and Wilsells, Uppsala, 438 pp.

Smith, C.S., 1961. Radio carbon dates from Easter Island. pp. 393-396, In: Heyerdahl, T. and Ferdon, E.N. (Eds). "Reports of the Norwegian

Archaeological Expedition to Easter Island and the East Pacific", Vol. 1, "Archaeology of Easter Island". Forum Publishing House, Stockholm, 559 pp.

Soil Conservation Service, 1975. "Soil Taxonomy". U.S. Department of Agriculture, Agriculture Handbook Number 436. U.S. Government Printing Office, Washington, D.C.

Steadman, D.W., Casanova, P. V., and C.C. Ferrando. (1994). Stratigraphy,

chronology, and cultural context of an early faunal assemblage from Easter Island. *Asian Perspectives* 33, 79-96.

Steadman, D.W. (1995). Prehistoric extinctions of Pacific Island birds: biodiversity meets zooarchaeology. *Science* 267, 1123-1131.

Stevenson, C.M., Wozniak, J., and Haoa, S. 1999. Prehistoric agricultural production on Easter Island (Rapa Nui), Chile. *Antiquity* 73, 801-812.

Stretten, N.A. and Zillman, J.W. (1984). Climate of the South Pacific Ocean. pp. 263-430 In: "Climates of the Oceans" (H. Van Loon, Ed). Volume 15 In: "World Survey of Climatology" (H.E. Landsberg, Editor in Chief). Elsevier, New York.

Stuiver, M., Reimer, P.J., Bard, E., Beck, J.W., Burr, G.S., Hughen, K.A., Kromer, B., McCormac, F.G., v. d. Plicht, J., and Spurk, M. (1998). INTCAL98 Radiocarbon age calibration 24,000 - 0 cal BP. Radiocarbon 40,1041-1083.

Wilmshurst, J.M. (1997). The impact of human settlement on vegetation and stability in Hawke's Bay, New Zealand. New Zealand Journal of Botany 35, 97-111.

Wozniak, J.A. (1998). Settlement patterns and subsistence on the northwest coast of Rapa Nui. pp. 185-192, In: "Easter Island in Pacific Context South Seas Symposium. Proceedings of the Fourth International Conference on Easter Island and East Polynesia". University of New Mexico, Albuquerque, 5-10 August 1997. C.M. Stevenson, G. Lee, and

F.J. Morin, Eds. The Easter Island Foundation, Bearsville and Cloud Mountain Press, Los Osos, CA.

Wright, C.S. and Diaz, V. (1962). "Soils and agricultural development of Easter Island (Pascua), Chile." Quarterly Report Supplement Number 1. Ministry of Agriculture, Santiago, Chile.

Yen, D.E. (1974). "The Sweet Potato and Oceania: An Essay in Ethnobotany". Bernice P. Bishop Museum. Bulletin 236.

Yen, D.E. (1990). Environment, agriculture and the colonization of the Pacific. pp. 258-277 In: "Pacific Production Systems: Approaches to Economic Prehistory." (D.E. Yen and J.M.J. Mummery, Eds). Papers from a Symposium at the XV Pacific Science Congress, Dunedin, New Zealand, 1983. Department of Prehistory, Research School of Pacific Studies, The Australian National University, Canberra.

Chapter 9

The Endemic Marine Invertebrates of Easter Island: How Many Species and for How Long?

CHRISTOPHER B. BOYKO
Division of Invertebrate Zoology, American Museum of Natural History, Central Park West @ 79th Street, New York, NY 10024 and Department of Biological Sciences, University of Rhode Island, Kingston, RI 02881

1. INTRODUCTION

By whatever name it is called, Easter Island, Isla de Pascua, or Rapa Nui, this small (ca. 106 km^2) spot of land in the central Pacific Ocean (27° 08'S, 109°20'W) has the distinction of being perhaps the most isolated spot on earth. It is approximately 3800 km from the South American mainland to the east, and over 2200 km from it's nearest neighbor to the west, Pitcairn Island. The closest landmass is tiny Sala y Gómez 415 km to the east, a bleak uninhabited rock. Easter Island is, of course, most famous for its anthropological history, and the numerous stone statues (moai) that dot the landscape. Not surprisingly, most of the preservation effort directed at Easter Island in this century has been towards these archaeological and cultural artifacts. That the biological component of the island has received less attention is also not surprising, given the relatively depauperate fauna and flora reported on and around the island. What animals lived on the island in historical times is unknown and the few terrestrial species living there today, from the common isopods to the ubiquitous hawks, were introduced either on purpose (hawks from Chile for rodent control; see Klemmer and Zizka, 1993) or by accident (cosmopolitan isopod species brought in on plants). Of the terrestrial invertebrate fauna, only the insects

Easter Island, Edited by John Loret and John T. Tanacredi
Kluwer Academic/Plenum Publishers, New York, 2003

have been studied in any detail (Campos & Peña, 1973); there are no comprehensive published accounts of the sparse isopod, mollusk, or flatwork faunas to date (see Kuschel, 1963; Klemmer and Zizka, 1993). The freshwater fauna contains no more unique species than does the land, being composed of only a few cosmopolitan rotifers (Segers & Dumont, 1993), microcrustaceans (Dumont & Martens, 1996), a dragonfly (Campos & Peña, 1973), and a small fish (*Gambusia* sp.) which was deliberately introduced in the 1930s, possibly for mosquito control.

There has been some widely scattered research done in the last 100 years on the marine fauna of Easter Island, most of it taxonomic in nature, although a few ecological studies have been made as well (e.g., Kohn, 1978; Orosio & Atan, 1993; Osorio & Cantuarias, 1989; Osorio *et al.*, 1999). The majority of previous investigations on the marine animals have been directed at the approximately 160 fish species (DiSalvo *et al.*, 1988). There have also been a few papers on annelids (Kohn & Lloyd, 1973; Rozbaczylo & Castilla, 1988), echinoderms (Fell, 1974; Massin, 1996), crustaceans (Holthuis, 1972; Garth, 1973; Garth, 1985; Fransen, 1987), and one major publication on the mollusks (Rehder, 1980). Most of the other marine invertebrates have either received little attention (e.g., Bryozoa, see Moyano, 1991; Porifera, see Desqueyroux-Faúndez, 1990) or none at all (e.g., Hemichordata, Nemertea [1], Sipuncula, Turbellaria). All of the published work has inevitably reached the same conclusion: the marine fauna of Easter Island, including fish, is depauperate compared to the other islands of the Indo-Pacific (Massin, 1996; Rehder, 1980) but harbors a high number of endemic taxa and is considered its own biogeographical province along with Sala y Gómez (or part of a province including Pitcairn Island to the west) (Briggs, 1974; Rehder, 1980).

But precisely how many endemic marine invertebrates does Easter Island harbor?

Given the limited samples of species that most previous researchers have reported, it has been difficult to estimate an overall value. Of the 26 species of brachyuran crabs reported in the literature from Easter Island, 15 are known from only a single specimen (Garth, 1973; 1985). Similar sampling problems exist with shrimp (Fransen, 1987; Holthuis, 1972), polychaetes (Kohn & Lloyd, 1973), and other smaller animals. In fact, only the mollusks appear to have been well sampled, and even in this phylum new species and new Easter Island records continue to be reported (e.g., Orosio, 1995; Poizat & Osorio, 1991). It appears that the level of endemicity for Easter Island marine invertebrates is very high (20-30%), but different levels of

[1] The first nemertean species has now been identified from Easter Island (see Boyko C.B. 2001. First Record of *Baseodiscus hemprichii* (Nemertea: Baseodiscidae) on Easter Island (Rapa Nui) and a New Eastern Distribution Boundary for the Species. Pacific Science 55(1): 41-42).

information available for different phyla make comparisons between groups difficult. For example, there are still no identified species of pycnogonids, sipunculans, nemerteans, or tunicates listed in the literature, yet specimens of all these groups have been collected from the island (DiSalvo *et al.*, 1988; Boyko, pers. obs.). Marine representatives of only eight phyla have been identified to species-level in published reports (see Table 1). There may well be additional identified specimens of other phyla represented in museum collections throughout the world, but as those data are not published, they cannot be used in analyses.

The following is a summary of the published data on the endemic species of marine invertebrates present on Easter Island. A brief discussion of the nature of the Rapanuian faunal province and its unique place in the zoogeographic framework is followed by an attempt to estimate the number of endemic species, both overall and in individual phyla, that are present on Easter Island today. Comparisons between the levels of endemism for marine invertebrates found on Easter Island and those of Hawaii and the Galápagos Islands are made, with notations of similarities and differences. This is followed by consideration of the factors impacting the endemic species of Easter Island, and how changing socio-political factors, such as the drive to increase tourism, may have a profound effect on the future survival of those taxa found only at the *Te Pito 'o te Henua* ("The End of the Land," an early Rapa Nui name for Easter Island).

2. THE RAPANUIAN FAUNAL PROVINCE

The idea of a unique faunal province for Easter Island originated with Schilder (1965), who used data obtained exclusively from studies of cypraeid mollusks to propose a series of biogeographical provinces worldwide. Schilder (1965) did not include Sala y Gómez in his original concept of the Rapanuian province, but later work incorporating additional taxa (e.g., Rehder, 1980) has shown that the two islands have much the same ecology and fauna, although the fauna of Sala y Gómez is still much more poorly known. The larger question is whether or not other islands to the west of Easter Island also belong to the Rapanuian faunal province (= "la provincia Pascuense" of Moyano, 1991). As noted by Rehder (1980), Easter Island has a strong biogeographic affinity with Hawaii (Hawaiian Province) to the northwest and also to Pitcairn Island (Polynesian Province) and the Kermadec Islands to the west. However, strong affinities between the faunas of these islands have only been well-documented for the Mollusca (Rehder, 1980). For other phyla with species having restricted (but non-endemic)

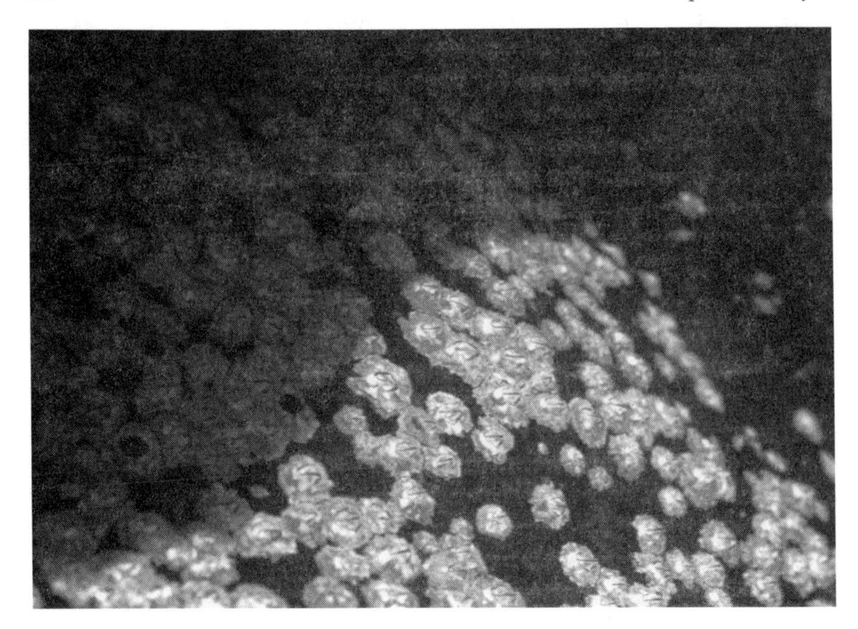

Figure 1. Rehderella belyaevi, the common barnacle on Easter Island; also found on Pitcairn Island. Photographed at Anakena.

distributions, the strongest affinities seem to be between Easter Island and Pitcairn Island, with several species known from only these two islands (e.g., the isopod *Cyathura rapanuia*, spiny lobster *Panulirus pascuensis* and barnacle *Rehderella belyaevi* (Fig. 1)). There is very little influence on the marine fauna of Easter Island from the South American mainland and all the species which these two areas share in common are broadly distributed circumtropical taxa (e.g., the seahare *Dolabella auricularia* and the shore crab *Leptograpsus variegatus).*

Some of the rarest invertebrate taxa found on Easter Island are apparently identical with broadly distributed Indo-Pacific taxa but have only been collected as single specimens on the island. Although some authors have suggested that these species occur as chance dispersal from western populations (DiSalvo *et al.*, 1988), the data available on the prevailing unidirectional east-west current flows appear to refute the likelihood of this happening on a frequent basis (Massin, 1996; Rehder, 1980). More probably, these "rare" taxa of Easter Island are actually living in low-density established populations, or in previously unexplored microhabitats. Some species that were formerly known from single, and presumed "waif" specimens, have been recollected years after the original record. An example of this is the wide ranging Indo-West Pacific starfish *Astropecten polyacanthus* (Fig. 2), formerly known on Easter Island from a single

Figure 2. Astropecten polyacanthus, an uncommon seastar previous known from a single specimen at Easter Island. This specimen collected at 97 feet depth, offshore from 'Ana 'o Keke.

specimen collected before 1988 (DiSalvo *et al.*, 1988), but found again there in 1999. Knowledge of a species' habitat is important before passing judgment on the relative scarcity of that taxon. For example, the endemic cowrie *Erosaria* (= *Cypraea*) *englerti* (Fig.3) was known from only 5 specimens before 1972. Once its nocturnal habits were discovered, it was readily collected in quantity from select tide pools (Rehder, 1980), although it is nowhere as common as the endemic and diurnal cowrie, *Erosaria* (= *Cypraea*) *caputdraconis*.

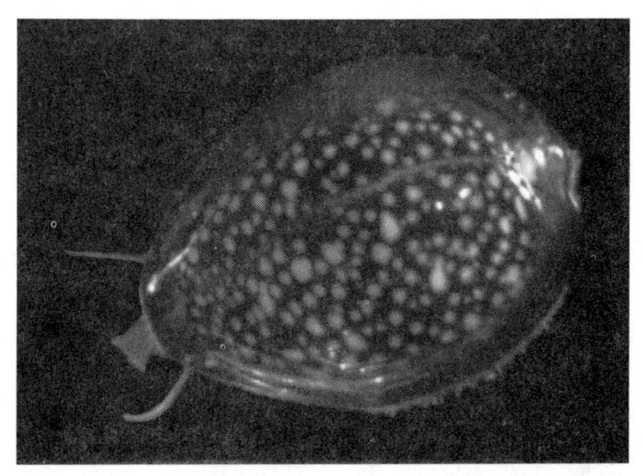

Figure 3. Erosaria englerti, an endemic, nocturnal and moderately common cowrie in tide pools. This specimen collected from Motu Marotiri, south shore of Poike.

Perhaps the most interesting thing about the endemic marine invertebrates of Easter Island is that they show a curious hybrid of tropical affinity coupled with temperate diversity. In tropical marine habitats (e.g., Hawaii), intertidal species diversity is typically high with relatively low numbers of any given taxon, while in temperate climates the marine intertidal biodiversity is low with high densities of a few taxa (Briggs, 1974).

On Easter Island, the marine fauna show clear biogeographic and taxonomic affinities to Indo-Pacific tropical taxa and little to no relationship with eastern Pacific fauna, yet these tropical taxa have evolved a temperate form of community structure, perhaps in response to the cool water temperatures (ca. 17.5-24° C [DiSalvo *et al.*, 1988]) found throughout the year around the island.

The intertidal zone on the island is primarily a region of volcanic rock (Fig. 4), with only a few small sandy beaches (Fig. 5). This zone is dominated by large numbers of very few endemic taxa, including the mollusks *Plaxiphora mercatoris* (local name "máma") (Fig. 6), *Erosaria caputdraconis* (local name "puré") (Fig. 7), *Nerita lirellata (local name "pipi uri"), Planaxis akuana, Pascula citrica, Siphonaria pascua, Conus miliaris pascuensis*, hermit crab *Calcinus pascuensis* sea urchin *Echinometra insularis* (local name "hatuke"), and the anemones *Actinogeton rapanuiensis* and *Zoanthus rapanuiensis* (both known locally as "takatore"). A similar

Figure 4. The sandy beach at Anakena. This strip of beach, the longest on the island, is considered the landing site of the King Hotu Matua, the first Polynesian to settle on Easter Island.

Figure 5. Near Tongariki, a typical intertidal region of volcanic rock with heavy surf action.

Figure 6. The common endemic chiton on Easter Island, *Plaxiphora mercatoris* is often found in sea urchin boreholes, with or without urchins. This specimen collected at Anakena.

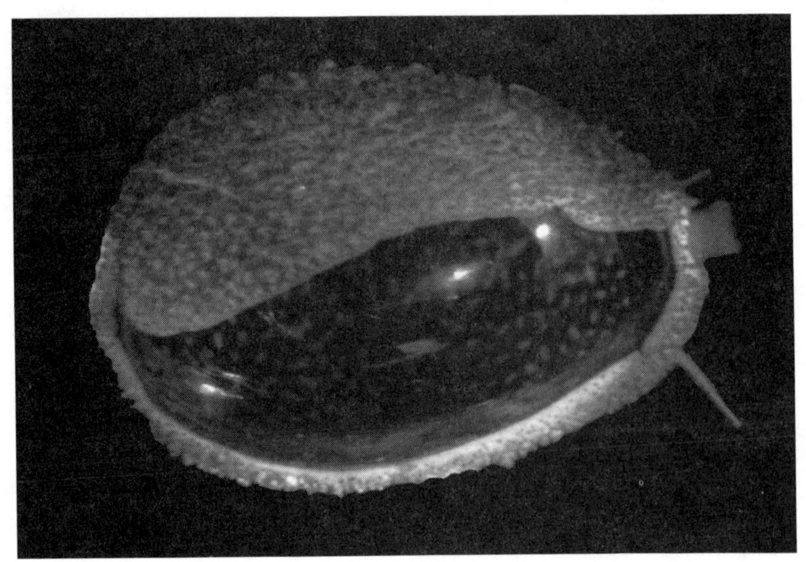

Figure 7. The endemic "serpent's head cowrie," *Erosaria caputdraconis*, is the most common cowrie on the island and is heavily collected for use in jewelry making by the islanders. This specimen collected from Motu Marotiri, south shore of Poike.

high concentration of shallow water endemic species is also found in the Hawaiian Islands (Kay, 1977), but overall biodiversity is much higher. The soft volcanic rock at the shoreline is pitted with innumerable holes made by *E. insularis,* and several other common endemic species (e.g., *E. caputdraconis* and *P. mercatoris*) use empty urchin boreholes as shelter from the pounding surf (Fig. 8). It has been suggested that the holes are actually made by chitons (Martin and Poppe, 1989), but the presence of holes exactly the size of the largest urchins and three times as big as the largest chiton belies this statement. Although the chitons may be contributors, it is the urchins who are the primary bioeroders on Easter Island.

Of the non-endemic species, the common intertidal Easter Island barnacle (*Rehderella belyaevi*) and littorinid snail *(Nodilittorina pyramidalis pascua*) also occur on Pitcairn Island. The few common taxa which are also broadly distributed in the Indo-Pacific include the sea cucumbers *Holothuria difficilis* and *H. cinerascens* (both known locally as "otake"), the mollusks *Nerita morio* and *Strombus maculatus* (Fig. 9), the hermit crab *Calcinus imperialis*, porcelain crab *Petrolisthes extremus*, shore crab *Leptograpsus variegatus* (known locally as "pikea", a name used for all brachyuran crabs), and fireworm *Eurythoe complanata.*

Figure 8. Sea urchin, *Echinometra insularis*, boreholes in volcanic rock at Anakena Beach. The urchins are endemic, as are many of the animals that share the boreholes with them.

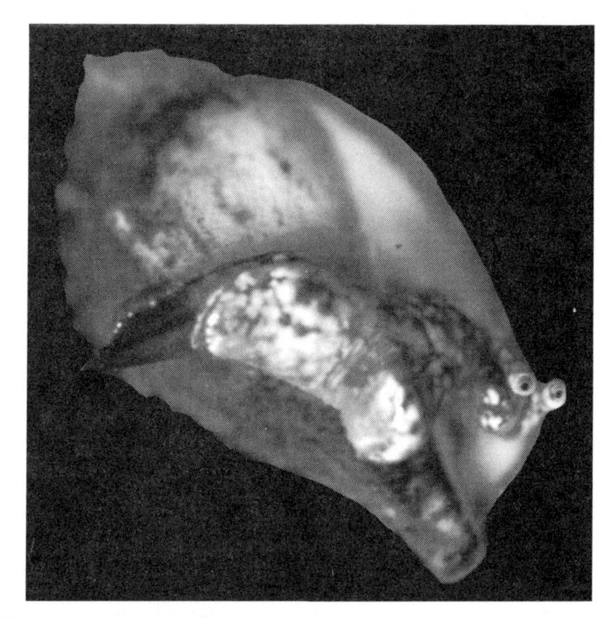

Figure 9. The only conch on Easter Island, *Strombus maculatus* is also widely distributed in the Indo-Pacific. This specimen collected at Anakena.

3. HOW MANY ENDEMICS?

The table below summarizes all known published information on the identified marine invertebrate species from Easter Island. Taxa without identified species on Easter Island are not included. Specimens identified only to family or genus, or as "sp." in the literature are not included, as their potential endemic nature cannot be determined.

Table 1. Biodiversity and Percentages of Endemic Marine Invertebrate Taxa on Easter Island

Taxon	Total # of Species	% Endemic	References
Annelida	43	2%	2, 12, 22
Arthropoda	75	20%	
Copepoda	7	14%	5, 26, 27
Ostracoda[2]	1	0%	28
Cirripedia	4	25%	6
Isopoda[3]	1	0%	1
Amphipoda	2	0%	4
Stomatopoda	1	0%	4
Decapoda	59	22%	4,7,8,9,10, 11,13,15,23
Bryozoa	23	17%	16
Cnidaria	19	32%	4
Echinodermata	26	12%	4,14
Mollusca	139	36%	4,17,18,19, 20,21,24
Platyhelminthes	6	0%	25,27
Porifera	18	39%	3

(1) Botosaneanu, 1987; (2) Cañete, 1997; (3) Desqueyroux-Faúndez, 1990; (4) DiSalvo *et al.*, 1988; (5) Fernández & Villalba, 1986; (6) Foster & Newman, 1987; (7) Fransen, 1987; (8) Garth, 1973; (9) Garth, 1985; (10) Haig, 1974; (11) Holthuis, 1972; (12) Kohn & Lloyd, 1973; (13) Kropp & Haig, 1994; (14) Massin, 1996; (15) McLaughlin & Haig, 1989; (16) Moyano, 1991; (17) Osorio, 1991; (18) Osorio, 1995; (19) Poizat & Osorio, 1991; (20) Prado, 1983; (21) Rehder, 1980; (22) Rozbaczylo & Castilla, 1988; (23) Saavedra *et al.*, 1996; (24) Senders & Martin, 1987; (25) Stunkard, 1965; (26) Villalba, 1987; (27) Villalba & Fernandez, 1985; (28) Wouters, 1997.

From the above data, it is clear that even among the taxa that are reasonably well documented from Easter Island, more comprehensive taxonomic work is needed, especially in those taxa which are both understudied and/or difficult to identify (e.g., Annelida, Amphipoda, Platyhelminthes). Future work in poorly studied groups with direct-developing young and low dispersal potential (e.g., amphipods, isopods) will no doubt increase their numbers of endemic taxa (see Barnard, 1991). Even

[2] A new publication on the Ostracoda of Easter Island puts the number of species at 31, with an endemicity of 81% (see Whatley, R., Jones, R. & Wouters, K. 2000. The marine Ostracoda of Easter Island. Rev. Española Micropaleontologia 32(1): 79-106).

[3] Preliminary results of examination of the isopod material collected in 1998/99 shows that at least 14 species are present on Easter Island, with an endemicity of approximately 86% (Brian Kensley, pers. comm.).

among phyla that are well documented (Mollusca, Porifera, Echinodermata), there are undoubtedly still more taxa awaiting discovery in the Easter Island biota. This is especially true among the microfauna which have a high potential for endemicity, as well as among taxa in the relatively unexplored subtidal regions.

The percentages of endemic taxa on Easter Island can be compared to similar faunas on other isolated islands in the Pacific. Newman and Foster (1983) pointed out that the relative youth (ca. 2.5 million years) and small size of Easter Island and Sala y Gómez theoretically work against the establishment of large numbers of endemic taxa in such a short span of time due to the expected high extinction rates in such a small geographic area. Although much larger in overall area, the Hawaiian (16,760 km^2) and Galápagos Islands (7845 km^2) are also relatively young landmasses which are known to support large numbers of endemic taxa. However, both the Hawaiian and Galápagos Islands appear to possess higher species diversity coupled with similar or even lower percentages of endemic taxa than Easter Island and Sala y Gómez (Table 2). This appears to be a consistent phenomenon found in all phyla, as can be seen by comparison with Table 1. Note that additional species in other phyla have been identified from both Hawaii and the Galápagos, but only the phyla that have been identified from these islands as well as Easter Island are included in Table 2.

From this data, it is evident that Easter Island either has far fewer taxa than Hawaii or the Galápagos, or is severely undersampled. Personal observations, and collection of over 4000+ specimens from Easter Island in 1998 and 1999, confirm that the reported low species numbers on Easter Island are an underestimate of the true diversity of the island, but accurately reflect the reality of its depauperate fauna compared to other areas in the Indo-Pacific, and even the tropical eastern pacific. As an example, from preliminary data given by DiSalvo *et al.* (1988) and the 1998/1999 collections, we know that there are perhaps twice as many species of decapod crustaceans on Easter Island than have been listed in the literature (Fig. 10). Conversely, intensive sampling has failed to increase the number of echinoderms known from Easter Island, which stand at a mere 9% of the total number of species found in Hawaiian waters. It appears that all phyla are more poorly represented at Easter Island than elsewhere in the tropical Pacific, but that some phyla are particularly depauperate.

In spite of the markedly lower biodiversity of Easter Island, a comparison of the percentage of endemic taxa there and in Hawaii and the Galápagos shows little substantial difference between these islands (where data are comparable.) Hawaii shows the greatest levels of endemism (22-≈95%), the Galápagos has a lower but still respectable range of percentages (15-38%), and Easter Island shows a similar range (2-39%). Clearly, a low level of biodiversity has not prevented the evolution of numerous unique species in the Easter Island environs to a level comparable with other Pacific Islands. More work on those groups currently having 0% endemism (e.g., isopods) is expected to raise these percentages considerably, as new species unique to Easter Island are in the process of being described at this time, based on the 1998/1999 material. The overall percentages of all three island regions indicate that all are deserving of the unique biogeographical province status they have been granted (Schilder, 1965; Briggs, 1974).

Table 2. Biodiversity and Percentages of Endemic Marine Invertebrate Taxa in Hawaii and the Galápagos Islands

Phylum/Class/ Order	Easter Island # spp. (% endemic)	Hawaii # spp. (% endemic)	Galapagos # spp. (% endemic)	Reference for Hawaii and Galápagos
Annelida	43 (2%)	281 (28%)	192 (31%)	6;4
Arthropoda	75 (20%)	?	?	
Copepoda	7 (14%)	100 (?%)	?	6
Ostracoda[4]	1 (0%)	79 (≈95%)	?	6, Eldredge, pers. comm.
Cirripedia	4 (25%)	49 (10%)	18 (22%)	Newman, pers. comm.; 13
Isopoda[5]	1 (0%)	27 (67%)	24 (33%)	6, Eldredge, pers. comm.; 5
Amphipoda	2 (0%)	120 (50%)	50 (38%)	2;3
Stomatopoda	1 (0%)	17 (65%)	4 (0%)	6; Hickman, pers. comm.
Decapoda	59 (22%)	438 (?%)	185 (15%)	6; 7, 12
Bryozoa	23 (17%)	150 (?%)	184 (18%)	6;1
Cnidaria	19 (32%)	339 (22%)	44 (20%)	6;11
Echinodermata	26 (12%)	278+ (54%)	198 (17%)	6,9;10
Mollusca	139 (36%)	787 (24%)	666 (18%)	6;8
Platyhelminthes	6 (0%)	407 (76%)	?	6
Porifera	18 (39%)	84 (29%)	?	6

(1) Banta, 1991; (2) Barnard, 1970; (3) Barnard, 1991; (4) Blake, 1991; (5) Brusca, 1987; (6) Eldredge & Miller, 1995; (7) Garth, 1991; (8) Kay, 1991; (9) Mah, 1998; (10) Maluf, 1991; (11) Wells, 1983 [hermatypic & ahermatypic corals only]; (12) Wicksten, 1991; (13) Zullo, 1991.

[4] See footnote 2. The current estimate of Easter Island ostracod diversity and endemicity, although higher than previously known, is still lower than that of Hawaii.

[5] See footnote 3. The level of Easter Island isopod diversity is still much lower than that of Hawaii and the Galápagos, but the endemicity appears considerably higher than at either of those locations.

Figure 10. Examples of crab species previously unreported from Easter Island. A) Dynomenidae sp., found both offshore and under rocks at Anakena; B) Majidae sp., from Anakena, male on the right, female on the left; C) Portunidae sp., from sandy bottom of a sponge covered cave in 40 feet depth, offshore of Hanga 'o Teo.

4. FOR HOW LONG?

Perhaps the greatest threat to the insular marine biota of Easter Island is the increasing pressure exerted on taxa impacted by the growing tourism industry. The most obviously affected species are those which are harvested for consumption by tourists and served in restaurants on the island. The main food invertebrate served to tourists is the local spiny lobster (*Panulirus pascuensis,* Fig. 11), whose populations have been severely reduced in number in the last few years (DiSalvo & Randall, 1993). This species is a traditional Rapanui food item, with the historical hunting of lobsters represented in numerous petroglyphs (Fig.12). This is a large species, previously reported with males reaching 25 cm total length and females growing to 24 cm total length (Holthuis, 1991), but males up to 39.5 cm total

Figure 11. The Easter Island spiny lobster, *Panulirus pascuensis*. Two large males (36.5 and 39.5 cm total length), a subadult male (15.1 cm total length) and a subadult female (17.0 cm total length). The 39.5 cm specimen is the largest example of the species recorded to date.

Figure 12. Petroglyphs from 'Ana 'o Keke, the so-called "Cave of the Virgins" on the Poike Volcano showing two different spiny lobster images.

length have been subsequently recorded (Boyko, pers. obs.). Surprisingly, spiny lobsters that grow above a certain weight (ca. 1 kg), may be partially protected from harvesting, as they become too large for the fishermen to easily sell them to restaurants (Boyko, pers. obs). To a lesser extent, the two endemic lobsters (*Parribacus perlatus* (Fig. 13) and *Scyllarides roggeveeni*) are also caught and sold to restaurants in quantity, but the annual catch of these two species has increased as the populations of spiny lobster have declined. Those endemic taxa which are collected by the native Rapanui for personal consumption (e.g., the urchin *Echinometra insularis*) are probably in no danger of overexploitation, unless an export fishery to the Asian market is implemented. The Easter Island octopus (*Octopus rapanui*) is another historically important food item for the Rapanui, and one that also appeals to Chileans living on Easter Island. This octopus may be strongly impacted by overfishing, especially with recent increased emigration to the island from Chile (see Porteous, 1993, for an overview of postcontact history).

One aspect of increased tourism that is not often addressed is the collection of local shellfish species for use in jewelry sold in the markets. Many of the species used by the local artisans for jewelry are actually Philippine or Tahitan imports. Two of the common mollusks seen in Rapanui jewelry are olive and cockle shells; neither of these two families occur on Easter Island. However, the most common shell species used in the making of jewelry are still local ones. Although some local shells used

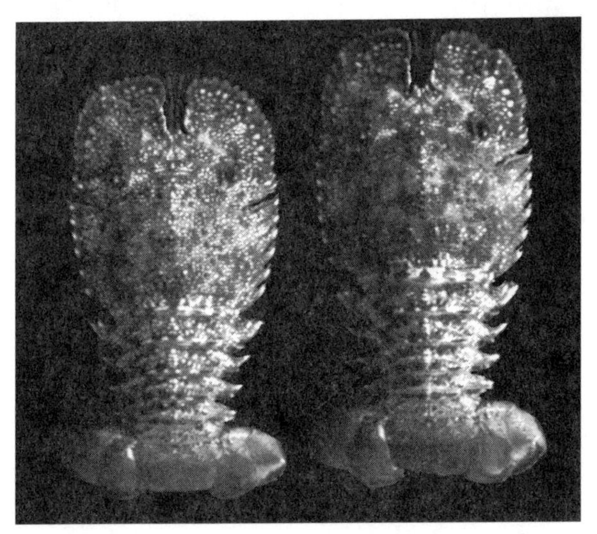

Figure 13. The endemic slipper lobster, *Parribacus perlatus*, is also the target of a fishery that caters to the island's restaurants. These specimens collected from Ovahe.

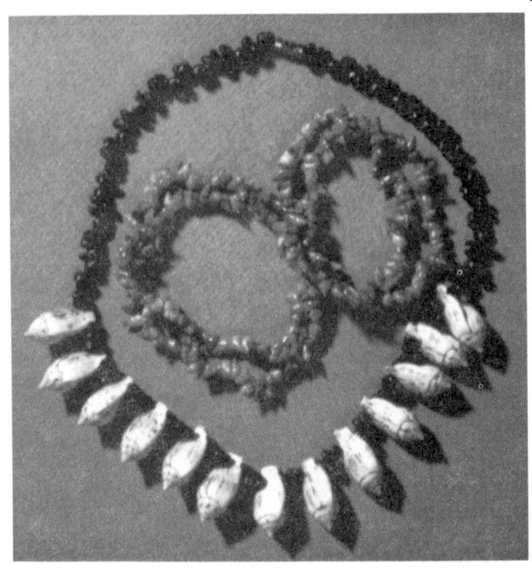

Figure 14. Two examples of the kinds of necklaces available for purchase by tourists in the marketplaces on Easter Island. The center necklace is made entirely of an endemic snail, *Planaxis akuana*, while the outer necklace is a mix of *Strombus maculatus* and *Nerita morio*, both wide-ranging Indo-Pacific taxa.

represent more broadly distributed Indo-Pacific taxa (*Strombus maculatus Nerita morio*), many of the most popular are endemic to Easter Island (e.g., *Erosaria caputdraconis, Planaxis akuana)* (Fig. 14). One species, the (non-endemic) sea cucumber parasite *Melanella cumingi* (Fig. 15), was formerly popular as a jewelry item but is now so scarce that it was not seen in any of the jewelry for sale during August 1999, and I found only 10 animals during almost two weeks of collecting, in spite of the host species being one of the two most common sea cucumbers on the island (see also DiSalvo *et al.*, 1988). Concerns about overharvesting of *Erosaria caputdraconis* have lead to suggestions of fishery moratoriums, at least for part of the year (Osorio *et al.*, 1999).

If plans to build a deep-water port for cruise ships on the island are realized, the increased pressure on the food species and the jewelry species will be further increased. Because a large proportion of the harvested taxa are endemics, and also among the most common taxa on the island, this could have a powerful impact on the biological diversity and stability of the island's marine habitat.

Figure 15. The sea cucumber parasite, *Melanella cumingi*, collected from Anakena on sea cucumbers. The shells were formerly common and used in jewelry making but are now scarce.

5. CONCLUSIONS

The marine invertebrate fauna of Easter Island has received too little attention for too long. Although it is markedly less diverse than those faunas of other tropical Pacific Islands, it contains a remarkably high percentage of endemic species for such a small and geologically young landmass. It has long been realized that the unique and highly specialized marine communities of Hawaii and the Galápagos Islands are in need of protection from overfishing and exploitation if they are to survive reasonably intact into the future. The time has come to recognize the special nature of Easter Island's marine community with its many endemic and threatened species. Although the native Rapanui peoples were historically responsible for the elimination of most native land fauna and flora, they apparently managed to keep a relatively harmonious balance with the marine world. Future dramatic increases in tourism on Easter Island, with its associated fishing and collecting burdens on the marine life, may eventually create an undersea habitat that rivals the land for lack of native species. Caution must be taken to ensure that the needs of the tourist trade are well balanced against the needs of the marine fauna of Easter Island and that "The End of the Land" does not suffer from "the end of the water."

ACKNOWLEDGMENTS

I would like to extend my thanks to all of the participants on the Science Museum of Long Island Easter Island Expedition: Blaine Cliver, Dennis Hubbard, Ellen Marsh, Rick & Susan Reanier, Hank Tonnemacher, and especially John T. Tanacredi and John Loret of the Science Museum of Long Island. This work was funded through the auspices of the Division of Natural Resources, Dr. John T. Tanacredi (retired), The National Park Service and Gateway National Recreation Area, presently the chairman, Department of Earth and Marine Science at Dowling College, Oakdale, NY. My sincere appreciation to the staff of the Hotel Topara'a: Sergio Rapu, Cecilia Rapu, Sergio Lopez, and Sergio Lopez, Jr. for all their daily assistance in collecting and gastronomic investigations. Special thanks to Michel Garćia for assistance in subtidal collection. Lu Eldredge (Bishop Museum, Hawaii), Cleveland Hickman (Washington and Lee University), Chris Mah (California Academy of Sciences), and William Newman (Scripps Institution of Oceanography) pointed out numerous helpful sources for Hawaiian and Galápagos data and offered some extremely useful unpublished results as well. Support from the American Museum of Natural History aided in film processing. Figure 14 is by Steve Thurston (AMNH); all other photographs are by the author.

REFERENCES

Banta, W.C., 1991, The Bryozoa of the Galápagos, pp. 371-389. In: James, M.J. (ed.). Galápagos Marine Invertebrates. Taxonomy, Biogeography, and Evolution in Darwin's Islands. Plenum Press, New York. 474 pp.

Barnard, J.L., 1970, Sublittoral Gammaridea (Amphipoda) of the Hawaiian Islands. Smithsonia Contr. Zoology 34: 286 pp.

Barnard, J.L., 1991, Amphipoda of the Galápagos Islands, pp. 193-206. In: James, M. J. (ed.). Galápagos Marine Invertebrates. Taxonomy, Biogeography, and Evolution in Darwin's Islands. Plenum Press, New York. 474 pp.

Blake, J.A., 1991, The polychaete fauna of the Galápagos Islands, pp. 75-96. In: James, M.J. (ed.) Galápagos Marine Invertebrates. Taxonomy, Biogeography, and Evolution in Darwin's Islands. Plenum Press, New York. 474 pp.

Botosaneanu, L., 1987, A new thalassostygobiont species of *Cyathura* (Isopoda: Anthuridea) from the south-east Pacific. Stygologia 3: 296-204.

Briggs, J.C., 1974, Marine Zoogeography. McGraw-Hill, New York. 475 pp.

Brusca, R.C., 1987, Biogeographic relationships of Galapagos marine isopod crustaceans. Bull. Mar. Sci. 41 (2): 268-281.

Campos, L., and Peña, L.E., 1973, Los insectos de Isla de Pasqua. Revue Chil. Ent. 7: 217-229.

Cañete, J.I., 1997, Descripción de cinco especies de Polynoidae (Polychaeta) de isla de Pascua. Rev. Biol. Mar. Oceanogr. 32: 189-202.

Castilla, J.C., and Rozbczylo, N., 1987, Invertebrados marinos de Isla de Pascua y Sala y Gómez. In: Islas Oceanicas Chilenas: Conocimiento Cientifico y Necesidades de Investigaciones. Ediciones Universidad Católica de Chile, Santiago. 356 pp.

Desqueyroux-Faúndez, R., 1990, Spongiaires (Demospongiae) de I'l'Ile de Pâques (Isla de Pascua). Revue Suisse Zool. 97: 373-409.

DiSalvo, L. H., and Randall, J.E., 1993, The marine fauna of Rapanui, past and present. pp. 16-23. In: Fischer, S.R. (ed.). Easter Island Studies. Contributions to the History of Rapanui in Memory of William T. Mulloy. Oxbow Monograph 32.

DiSalvo, L.H., Randall, J.E., and Cea, A., 1988, Ecological reconnaissance of the Easter Island sublittoral marine environment. Natl. Geogr. Res. 4: 451-473.

Dumont, H.J., and Martens, K., 1996, The freshwater microcrustacea of Easter Island. Hydrobiologia 325: 83-99.

Eldredge, L.G., and Miller, S.E., 1995, How many species are there in Hawaii? Bishop Mus. Occ. Pap. 41: 3-18.

Fell, F.J., 1974, The echinoids of Easter Island (Rapa Nui). Pac. Sci. 28: 147-158.

Fernández, J. and Villalba, C. 1986, Contribucion al conocimiento del genero *Caligus* Müller, 1785 (Copepoda: Siphonostomatoida) en Chile. Gayana Zool. 50: 37-62.

Foster, B.A., and Newman, W.A., 1987, Chthamalid barnacles of Easter Island; peripheral Pacific isolation of Notochthamalinae new subfamily and *hembeli*-group of Euraphiinae (Cirripedia: Chthamalidoidea). Bull. Mar. Sci. 41: 322-336.

Fransen, C.H.J.M., 1987, Notes on caridean shrimps of Easter Island with descriptions of three new species. Zool. Meded. 61: 501-531.

Garth, J.S., 1973, The brachyuran crabs of Easter Island. Proc. California Acad. Sci., 4th ser. 39: 311-336.

Garth, J.S., 1985, On a small collection of brachyuran Crustacea from Easter Island obtained by the Scripps Institution of Oceanography Downwind Expedition of 1958. Occ. Pap. Allan Hancock Found., n. ser. 3: 12 pp.

Garth, J.S., 1991, Taxonomy, distribution, and ecology of Galápagos Brachyura, pp. 123-145. In: James, M.J. (ed.). Galápagos Marine Invertebrates. Taxonomy, Biogeography, and Evolution in Darwin's Islands. Plenum Press, New York. 474 pp.

Haig, J., 1974, *Calcinus pascuenis*, a new hermit crab from Easter Island (Decapoda, Anomura, Diogenidae). Crustaceana 27: 27-30.

Holthuis, L.B., 1972, The Crustacea Decapoda Macrura (the Alpheidae excepted) of Easter Island. Zool. Meded. 476: 29-54, 2 pls.

Holthuis, L.B., 1991, Marine lobsters of the world. FAO Fisheries Synopsis 125 (13): 292 pp.

Kay, E.A., 1977, Introduction to the revised edition, pp. 4-11. In: Devaney, D.M. and Eldredge, L.G. (eds.). Reef and Shore Fauna of Hawaii. Section 1: Protozoa through Ctenophora. Bernice P. Bishop Mus. Spec. Publ. 64 (1).

Kay, E.A., 1991, The marine mollusks of the Galápagos determinants of insular marine faunas. pp. 235-252. In: James, M.J. (ed.). Galápagos Marine Invertebrates. Taxonomy, Biogeography, and Evolution in Darwin's Islands. Plenum Press, New York. 474 pp.

Klemmer, K., and Zizka, G., 1993, The terrestrial fauna of Easter Island, pp. 24-26. In: Fischer, S.R. (ed.). Easter Island Studies. Contributions to the History of Rapanui in Memory of William T. Mulloy. Oxbow Monograph 32.

Kohn, A.J., 1978, Gastropods as predators and prey at Easter Island. Pac. Sci. 32: 35-37.

Kohn, A.J., and Lloyd, M.C., 1973, Marine polychaete annelids of Easter Island. Int. Revue Ges. Hydrobiol. 58: 691-712.

Kropp, R.K., and Haig, J., 1994, *Petrolisthes extremus*, a new porcelain crab (Decapoda: Anomura: Porcellanidae) from the Indo-Pacific. Proc. Biol. Soc. Washington 107: 312-317.

Kuschel, G., 1963, Composition and relationship of the terrestrial faunas of Easter, Juan Fernandez, Desventuradas, and Galapágos Islands. Occ. Pap. California Acad. Sci. 44: 79-95.

Mah, C., 1998, new records, taxonomic notes, and a checklist of Hawaiian starfish. Bishop Mus. Occ. Pap. 55: 65-71.

Maluf, L.Y., 1991, Echinoderm fauna of the Galápagos Islands, pp. 345-367. In: James, M.J. (ed.). Galápagos Marine Invertebrates. Taxonomy, Biogeography, and Evolution in Darwin's Islands. Plenum Press, New York. 474 pp.

Martin, P. and Poppe, G.T., 1989, Notes on the Mollusca of Easter Island: *Cypraea*. Hawaiian Shell News 37(10): 1, 7.

Massin, C., 1996, The holothurians of Easter Island. Bull. Inst. Royal Sci. Nat. Belgique 66: 151-178.

McLaughlin, P.A., and Haig, J., 1989, On the status of *Pylopaguropsis zebra* (Henderson), *P. magnimanus* (Henderson), and *Galapagurus teevanus* Boone, with descriptions of seven new species of *Pylopaguropis* (Crustacea: Anomura: Paguridae). Micronesica 22: 123-171.

Moyano, H.I., 1991, Bryozoa marinos Chilenos VIII: una sintesis zoogeografica con consideraciones sistematicas y la description de diez especies y dos generos nuevos. Gayana Zool. 55: 305-389.

Newman, W.A., and Foster, B.A., 1983, The Rapanuian faunal district (Easter and Sala y Gómez): in search of ancient archipelagos. Bull. Mar. Sci. 33: 633-644.

Osorio, C., 1991, *Charonia tritonis* (Linne 1758) en Isla de Pascua (Mollusca: Gastropoda: Cymatiidae). Rev. Biol. Mar. Valparaíso 26: 75-80.

Orosio, C., 1995, Dos nuevos registros de Isognomiidae (Mollusca Bivalvia) para Isla de Pascua, Chile, Rev. Biol. Mar. Valparaíso 30: 199-205.

Osorio, C., and Atan, H., 1993, Relaciones biologias entre *Luetzonia goodingi* Rehder, 1980 (Gastropoda, Stiliferidae) parasito de *Echinometra insularis* Clark, 1972 [sic] (Echinoidea) en Isla de Pascua. Rev. Biol. Mar. Valparaíso 28: 99-109.

Osorio, C., and Cantuarias, V., 1989, Vertical distribution of the mollusks on the rocky intertidal of Easter Island. Pac. Sci. 43: 302-315.

Osorio, C., Brown, D., Donoso, L., and Atan, H., 1999, Aspects of the reproductive activity of *Cypraea caputdraconis* from Easter Island (Mollusca: Gastropoda: Cypraeidae). Pac. Sci. 53: 15-23.

Poizat, C.H., and Osorio, C., 1991, *Aplysia (Varria) dactylomela asymetrica*, sous-espece nouvelle (gasteropode, opisthobrache), a l'Ile de Pâques, Chili. Mésogee 51: 59-63.

Porteous, J.D., 1993, The modernization of Rapanui, pp. 225-227. In: Fischer, S.R. (ed.). Easter Island Studies. Contributions to the History of Rapanui in Memory of William T. Mulloy. Oxbow Monograph 32.

Prado, R., 1983, Nota sobre cefalopodos decapodos colectados alrededor de las Islas de Pascua y Sala y Gomez. Bol. Soc. Biol. Concepción 54: 159-162.

Rehder, H.A., 1980, The marine mollusks of Easter Island (Isla de Pascua) and Sala y Gómez. Smithsonian Contrib. Zool. 289: 167 pp.

Rozbaczylo, N., and Castilla, J.C., 1988, A new species of polychaeta, *Scololepis anakenae* (Polychaeta: Spionidae) from Easter Island, south Pacific Ocean, with ecological comments. Proc. Biol. Soc. Washington 101: 767-772.

Saavedra, M., Carvacho, A., and Letelier, S., 1996, Nuevo registro de *Metabetaeus minutus* (Whittelegge) (Crustacea, Decapoda, Alpheidae) en Isla de Pascua. Rev. Biol. Mar. Valparaíso 31: 117-122.

Schilder, F.A., 1965, The geographical distribution of cowries (Mollusca: Gastropoda). Veliger 7: 171-183.

Segers, H., and Dumont, H.J., 1993, Zoogeography of Pacific Ocean Islands: a comparison of the rotifer faunas of Easter Island and the Galápagos archipelago. Hydrobiologia 255/256: 475-480.

Senders, J., and Martin, P., 1987, Description d'une nouvelle sous-espece de Cypraeidae en provenance de l'Ile de Pâques. Apex 2: 13-24.

Stunkard, H. W., 1965, A digenetic trematode, *Botulus cablei,* n. sp., from the stomach of the lancetfish, *Alepisaurus borealis* Gill, taken in the south Pacific. Biol. Bull. Woods Hole 128: 488-491.

Villalba, C., 1987, Contribucion al concimiento del genero *Hatschekia* Poche, 1902 en Chile (Copepoda: Hatschekiidae). Bol. Soc. Biol. Concepción 57: 155-170.

Villalba, C., and Fernandez, J., 1985, Parasitos de *Mola ramsayi* (Giglioli, 1883) (Pices: Molidae) en Chile. Bol. Soc. Biol. Concepción 56: 71-78.

Wells, J.W., 1983, Annotated list of the scleractinian corals of the Galápagos. pp. 213-291. In: Glynn, P.W. and Wellington, G.M. (eds.). Corals and Coral Reefs of the Galápagos Islands. University of California Press, Berkeley. 330 pp.

Wicksten, M.K., 1991, Caridean and stenopodid shrimp of the Galápagos Islands, pp. 147-156. In: James, M.J. (ed.). Galápagos Marine Invertebrates. Taxonomy, Biogeography, and Evolution in Darwin's Islands. Plenum Press, New York. 474 pp.

Wouters, K., 1997, A new genus of the family Pontocyprididae (Crustacea, Ostracoda) from the Indian and Pacific Oceans, with the description of two new species. Bull. Inst. Royal Sci. Nat. Belgique 67: 67-76.

Zullo, V.A., 1991, Zoogeography of the shallow-water cirriped fauna of the Galápagos Islands and adjacent regions in the tropical eastern Pacific, pp. 173-192. In: James, M.J. (ed.). Galápagos Marine Invertebrates. Taxonomy, Biogeography, and Evolution in Darwin's Islands. Plenum Press, New York. 474 pp.

Chapter 10

Finfish in the Rano Kau Caldera of Easter Island

LUCIA MAGLIULO-CEPRIANO[1,2], MARTIN P. SCHREIBMAN[2], AND JOHN T. TANACREDI[3]
[1]Farmingdale State University of New York, Farmingdale, New York 111735; [2]Aquatic Research and Environmental Assessment Center (AREAC) at Brooklyn College, City University of New York 11210; Department Earth & Marine Science, Dowling College, Oakdale, New York

1. INTRODUCTION

The Rano Kau Caldera is the largest inactive volcano on Easter Island. This Chilean island, a triangle of volcanic rock located 2,600 miles west of Santiago, is the most remote of all populated islands on our planet. Rano Kau is located on the southwestern tip of the island. Its crater is the basin of a large freshwater lake. As part of the cooperative expedition undertaken by the National Park Service of the United States and CONAF of Chile, fish specimens were collected from the Rano Kau Caldera and brought back to the United States for identification and study.

2. COLLECTION AND PROCESSING OF SPECIMENS

The fish were collected on August 28, 1999, from the Rano Kau Caldera using a small seine. Whole specimens were immediately preserved and stored in a 10% formalin solution. In the laboratory, the specimens were rinsed, measured (standard length, from snout to caudal fin) and autopsied. Their stage of sexual development and the gross morphology of reproductive and visceral organs were evaluated at autopsy. Heads and bodies were

Easter Island, Edited by John Loret and John T. Tanacredi
Kluwer Academic/Plenum Publishers, New York, 2003

placed in Bouins solution, under vacuum for 24 hrs, then decalcified, dehydrated in alcohol, and embedded in wax (polyfin). Wax-embedded specimens were then cut into five micrometer thick slices which were mounted on gelatin-coated glass slides. Representative slides were stained with Masson's Trichrome while the remaining slides were processed for immunocytochemical (ICC) procedures utilizing the Elite ABC kit (Vector Laboratories, CA.). These procedures would allow for the histological and biochemical analysis of body tissues. All standard ICC controls, including the absorption of the primary antibody with its antigen and the substitution of normal serum for active reagents in subsequent steps of the ICC protocol, were carried out to insure the validity of our results.

3. IDENTIFICATION OF SPECIMENS

All of the fish were of a single species. They were identified as freshwater teleosts of the genus *Gambusia*. They were further distinguished as *G. affinis*, commonly known as the mosquitofish or the western mosquitofish, based on the presence of 7 rays in the dorsal fins and 10 in the anal fins of collected specimens (Walters and Freeman, 2000). The western mosquitofish, and their close relatives, the eastern mosquitofish, *G. holbrooki*, are common inhabitants of freshwater habitats throughout the southern and western United States and in many tropical and sub-tropical regions of the world. The diet of the mosquitofish tends to vary with the season or with availability. They tend to consume large quantities of algae in winter and spring and equally large quantities of insects, insect larvae, and small crustaceans in summer months (Gophen *et al.*, 1998). In recent years, mosquitofish have been introduced to a number of foreign habitats as fairly effective components of mosquito-control programs, particularly in regions where malaria is prevalent (Persichino *et al.*, 1998, Goodsell and Kats, 1999). The common name of these animals reflects their taste for mosquito larvae. The mosquitofish are viviparous teleosts; they are capable of internal fertilization and give birth to free-swimming young. They are sexually dimorphic; gender is easily determined based on the structure of the anal fin (see below).

4. AUTOPSY RESULTS

The sex and the stage of sexual development of the collected specimens was determined by visual examination of the gonads and of the anal fin. Reproductively immature fish of both sexes have a fan-shaped anal fin and an undifferentiated gonad that appears to be composed of clear translucent tissue to the naked eye. Females of the species undergo an annual cycle of ovarian development in which the ovary regresses to an immature stage during winter months and then undergoes recrudescence, a re-ripening of the ovary, in the spring. In the initial stages of recrudescence, as in the initial stages of sexual development, the ovary will be populated with small, white, non-yolky oocytes and some very young oocytes, said to be in the oil droplet stage. As the process continues, yolk is gradually deposited in the oocytes. Reproductively mature females will have large, yellow, yolky oocytes in the ovary (Koya, *et al.*, 1998; Koya, *et al.*, 2000).

In the male, the immature gonad remains small, clear and translucent until spermatogenesis begins. The onset of sperm production will cause the gonad to become larger, opaque and white. The anal fin of male mosquitofish, under the influence of androgens during sexual maturation, undergoes a transformation that converts the fin into an intromittent organ, known as a gonopodium, capable of internal fertilization during copulation (Batty and Lim, 1999). The gonopodium develops in males in conjunction with increasing secretion of androgens from the developing gonads.

Of the 32 specimens collected, 15 were identified as mature males, 10 were identified as immature females, and 8 animals had undifferentiated gonads. All 15 males had fully developed gonopodium and large, white opaque testes. The males ranged in length from 17.0 millimeters (mm) to 20.5 mm with average standard length equal to 19.0 mm. The 10 immature females were found to have small, white oocytes in their ovary. They ranged in length from 19.0 mm to 27.0 mm with the average standard length equal to 21.9 mm. The undifferentiated animals had fan-shaped anal fins and clear, translucent gonads. They ranged in length from 15.5 mm to 26.5 mm with the average standard length equal to 21.1 mm. In size, the undifferentiated animals resembled the females.

Mosquitofish are known to vary in size from 20 mm to 60 mm. The collected animals, therefore, were small for fish of this species. The size of the fish has been linked to their diet since the overall size of the animal will influence the gape size. Mosquitofish 25 mm or less are more likely to feed on zooplankton, such as small insects and insect larvae, whereas larger animals were more likely to feed on larger invertebrates (Mansfield and McArdle, 1998). The small size of these animals may therefore reflect the type of food that was most easily available to them. Thermal stability of the environment also appears to have a significant effect on the size of mosquitofish. Mosquitofish from thermally stable environments have been shown to mature at larger sizes, have higher fat reserves and larger embryos than those from thermally unstable environments (Stockwell and Vinyard, 2000).

In all other respects, the autopsy revealed that these animals were healthy at the time of collection.

5. HISTOLOGICAL EVALUATION

Histological evaluation of the gonads of males revealed fully mature testes containing advanced stages of spermatogenesis. The gonads of females revealed ovaries containing oocytes that had not yet begun yolk deposition. The gonads of the undifferentiated animals, which were indiscernible as either male or female at autopsy, were revealed to be immature ovaries. The oocytes in the ovaries appeared to be newly formed, some in the oil droplet stage, but all very immature.

Histological evaluation of livers, kidneys, digestive organs and brains confirmed our autopsy findings that these animals were in basic good health at the time of collection.

6. IMMUNOCYTOCHEMICAL EVALUATION

Immunocytochemistry (ICC) is a means of identifying and locating specific chemical components *(ie.*, hormones and neuropeptides) of cells and tissues by utilizing antibodies generated against the component. The particular ICC method utilized in this study was the avidin-biotin method (ABC Elite, Vector Lab., CA.). In order to evaluate the organization of the brain and the pituitary gland of the collected specimens, specific antibodies to a number of known neuroendocrine components were utilized. We investigated the distribution of the following compounds:

FMRF-amide: This tetrapeptide is known to be distributed in regions of the pre-optic area and the hypothalamus of the brain involved in the control of reproduction (Rama Krishna, *et al.*, 1992). It is specifically a marker for the nucleus olfactoretinalis (NOR; terminal nerve), a brain center believed to act as an integrator of internal and external cues that guide maturation of the reproductive system and the onset of reproductive system.

Gonadotropin Releasing Hormone (GnRH): This decapeptide is closely associated with the control of the production and release of the pituitary gonadotropic hormones. Its presence in NOR has been linked to the onset of reproductive maturation (Schreibman and Margolis-Nunno, 1987). In teleost fish, this hypothalamic protein is released in close proximity to pituitary cells responsible for the production and secretion of gonadotropins.

Gonadotropins (GTH): These glycoprotein hormones are produced by gonadotropic cells of the anterior pituitary gland. They are the main regulators of gonadal function. They function to maintain the gonad, regulate the timing of reproductive cycles and the maturation of the gametes. Sexually immature teleost fish have few GTH cells in the pituitary gland and thus produce little GTH. As the animal begins sexual maturity, under the influence of GnRH from the hypothalamus, the GTH cells along the medial ventral border of the pituitary glad begin to proliferate. As this proliferation proceeds, these cells will form a gonadotropic zone that girdles the ventral border of the gland. The cells in the gonadotropic zone become active and begin to produce and secrete significant quantities of GTH resulting in development and maturation of the gonad, followed by gametogenesis (Schreibman, 1986).

In the collected specimens, we found FMRF-amide in cells of the NOR, with traces of this compound extending posteriorly along neural tracts through the pre-optic area and into the hypothalamus, towards the pituitary gland. The distribution of FMRF-amide in all fish studied was in keeping with the distribution of this hormone in sexually mature teleosts (Magliulo-Cepriano, *et al.*, 1993).

GnRH was found in the hypothalamus and neurohypophysis of the pituitary gland of studied fish. Granules were also located around cells of the ventral anterior pituitary gland.

GTH was found in the cells of the ventral pituitary gland forming a full and active gonadotropic zone. This zone was found in all fish studied. The presence of this zone is an indicator of sexual maturity (Schreibman, 1986).

This ICC evidence indicates that all 32 of the collected specimens were mature fish. The distribution of all three compounds investigated was in keeping with what would be expected in mature fish. Animals that were thought to be immature or undifferentiated at autopsy were more likely to be in an earlier stage of ovarian recrudescence. This is supported by the presence of a fully active gonadotropic zone in the pituitary gland of these animals and by the histological evaluation of their gonad which showed the presence of immature, newly formed oocytes. It is further bolstered by the fact that the undifferentiated fish were essentially of the same size as the distinguishable females.

It is also in keeping with what is known of the reproductive cycles of these animals. Members of the genus *Gambusia* have been known to be continuous breeders in laboratory settings but in the wild they undergo reproductive cycles as described above. The specimens collected on Easter Island were collected in late August, which is early spring there. That is the time of the year when females would be expected to undergo ovarian recrudescence in preparation for the oncoming breeding season. Reproductive function in the female is closely linked to both temperature and photoperiod. Temperature appears to play the greater role in the onset of reproductive activities, with the development of mature oocytes and pregnancy during spring, while photoperiod appears to have a greater influence over the cessation of reproductive activities with the regression of the ovary in late summer (Koya and Kamiya, 2000).

7. CONCLUSIONS

The fish collected from the Rano Kau Caldera of Easter Island were healthy, adult western mosquitofish, presumably on the brink of a new breeding season. The probability is high that these animals were, at some time, introduced to the caldera as a means of mosquito control. These animals are hardy and aggressive. They have been known to be detrimental to endogenous species of fish, amphibians and, of course, insects (Komak and Crossland, 2000). Thus, it would be interesting to know if environmental impact studies were done to evaluate the effects of the introduction of this species to the caldera, if indeed, they were introduced, as suspected. While we cannot guess at the impact on the environment, as a result of this study, what is known is that these animals are currently thriving in the caldera. In addition, our preliminary observations suggest that further study is warranted to confirm that the developmental stage of the animals we studied is indeed season-related and that our observations are similar for mosquitofish in other caldera on Easter Island.

ACKNOWLEDGMENTS

The capable technical assistance of Ms. Nasiha Ocasio is deeply appreciated. This study was supported by a memorandum of agreement between AREAC and the U.S. National Park Service at Gateway. Ms. Ocasio was supported by the Minority Access to Research Careers (MARC) program at Brooklyn College, C.U.N.Y.

REFERENCES

Batty, J., and Lim, R., 1999, Morphological and reproductive characteristics of male mosquitofish (*Gambusia affinis* holbrooki) Inhabiting sewage-contaminated waters on New South Wales, Australia. *Arch. Environ. Contam. Toxicol.* 36, 301-307.

Goodsell, J.A., and Kats, L.B., 1999, Effects of introduced mosquitofish on pacific treefrogs and the role of alternative prey. *Conserv. Biol.* 13, 921-924.

Gophen, M., Yehuda, Y., Malinkov, A., Degani, G., 1998, Food composition of the fish community in Lake Agmon. *Hydrobiologia* 380, 49-57.

Komak, S., and Crossland, M.R., 2000, An assessment of the introduced mosquitofish (*Gambusia affinis holbrooki*) as a predator of eggs, hatchlings and tadpoles of native and non-native anurans. *Wildl. Res.* 27, 185-189.

Koya, Y., and Kamiya, E., 2000, Environmental regulation of annual reproductive cycle of the mosquitofish, *Gambusia affinis*. *J. Exp. Zool.* 286, 204-211.

Koya, Y., Inoue, M., Naruse, T., Sawaguchi, S., 2000, Dynamics of oocyte and embryonic development during ovarian cycle of the viviparous mosquitofish *Gambusia affinis*. *Fish. Sci.* 66, 63-70.

Koya, Y., Itazu, T., Inoue, M., 1998, Annual reproductive cycle based on histological changes in the ovary of the female mosquitofish, *Gambusia affinis*, in cental Japan. *Ichthyol. Res.* 45, 241-248.

Magliulo-Cepriano, L., Schreibman, M.P., Blum, V., 1993, The distribution of immunoreactive FMRF-amide, neurotensin and galanin in the brain and pituitary gland of three species of *Xiphophorus* from birth to sexual maturity. *Gen. Comp. Endocrinol.* 92, 269-280.

Mansfield, S., and McArdle, B.H., 1998, Dietary composition of *Gambusia affinis* (Family *Poeciliidae*) populations in the northern Waikato region of New Zealand. *N.Z.J. Mar. Freshwat. Res.* 32, 375-383.

Persichino, J., Christian, E., Carter, R., 1998, Short term effects of Bolero and the gill apparatus of a small number of mosquitofish (*Gambusia affinis*). *Bull. Environ. Contam. Toxicol.* 61, 162-168.

Rama Krishna, N.S., Subhedar, N., Schreibman, M.P., 1992, FMRF-amide like immunoreactive nervus terminalis innervation to the pituitary in the catfish, *Clarias batrachus*: Demonstration by lesion and immunocytochemical techniques. *Gen. Comp. Endocrinol.* 85, 111-117.

Schreibman, M.P., 1986, The pituitary gland. In: *Vertebrate Endocrinology: Fundamentals and Biomedical Implications. Vol. I*, Academic Press, pp. 11-55.

Schreibman, M.P., and Margolis-Nunno, H., 1987, Reproductive biology of the terminal nerve (nucleus olfactoretinalis) and other LHRH pathways in teleost fishes. *Ann. N.Y. Acad. Sci.* 519, 60-68.

Stockwell, C.A., and Vinyard, G.L., 2000, Life history variation in recently established populations of western mosquitofish (*Gambusia affinis*). *West. N. Am. Nat.* 60, 273-280.

Walters, D.M., and Freeman, B.J., 2000 Distribution of *Gambusia (Poeciliidae)* in a southeastern river system and the use of fin ray counts for species determination. *Copeia* 2, 555-559.

PART III

TECHNOLOGICAL APPLICATIONS TO PROTECT BIOLOGICAL AND CULTURAL ARTIFACTS

The largest and most impressive ahu on Easter Island. The largest moai here is well over 90 tons in weight.

Chapter 11

Aerial Surveys of Isle De Pasqua: Easter Island and the New Birdmen

ROBERT A. HEMM and MARCELO MENDEZ
Science Museum of Long Island and Explorer's Club

1. INTRODUCTION

Easter Island is one of the most isolated and fascinating locales on Earth. Its mysteries have spawned volumes of speculation and dozens of expeditions over two and a half centuries since its discovery by European Explorers in the early 18th century. Located more than 2000 miles west of South America and about the same distance from Tahiti to the west, the island rises abruptly from the sea and is pounded by the ocean on all sides. On only sixty-four square miles can be found the remains of a culture which produced many unique features including gigantic statues weighing scores of tons, large stone burial platforms, a profusion of petrographs, and a written language yet to be deciphered. Within a thousand years the people of Easter Island deforested the island, crowded many plants and animals to extinction, and saw their society descend into social and political chaos and even cannibalism. Easter Island's culture, much like that of the Mayans but existing on a far more limited and fragile land mass, may have fallen for many of the same reasons: overpopulation (with no other place to go), environmental degradation, warfare among the clans for diminishing resources and perhaps adverse weather conditions such as a prolonged El Nino which could have created excessive rains or drought.

Easter Island, Edited by John Loret and John T. Tanacredi
Kluwer Academic/Plenum Publishers, New York, 2003

Most of the immediate past expeditions to Easter Island were tied to Anthropology or Archaeology - trying to determine when and where the original inhabitants came from – perhaps Polynesia or South America? Or, perhaps both at different times? The Icon for Easter Island is the many massive statues carved and erected over a period of hundreds of years, estimated to be about 900 to 1400ce. How were the many hundreds of massive statues, weighing as much as 100 tons, carved and transported to sites around the island. And how were they erected, all by a people without the aid of draft animals, the wheel or metal tools?

The first modern full-scale expedition took place in 1955 led by Thor Heyerdahl. He carried out extensive work in archaeology and anthropology. He came by ship with over 35 scientists and crew and stayed over nine months. Thor carried with him Explorers Club Flag #123, the same flag he carried on his famous Kon Tiki Voyage in 1947.

2. OUR WORK

In 1997 a Scientific Interdisciplinary Expedition was formed under the Sponsorship of The Science Museum of LI and its Director, Dr. John Loret. An Oceanographer and three time President of the Explorers Club, Loret was a member of Heyerdahl's Expedition in 1955 and quite appropriately awarded Heyerdahl's Flag.

The primary objectives of the '97 and the '98 expedition which followed it was to gather a group of scientists in diverse fields including paleobotany, geology, marine biology and archaeology to try to answer some of the outstanding questions.

What did Easter Island look like before man first arrived sometime in the 4th century? What were the varieties of plant life, land animals, and birds? To find the answers, scientists drilled cores in the swampy peat of extinct volcanoes and were able to go down to 55 feet deep to measure plant history going back some 85,000 years.

Other scientists using specially developed coring drills were involved in diving on deep water corals surrounding the island. They were able to get sample cores some 5 feet in length. The purpose was determining past weather patterns over the past 500 years and answering the question – was the decline of Easter Island's civilization due only to man's overuse of his resources, or were there other factors, such as severe weather changes? Perhaps a prolonged El Nino?

As a co-leader with John Loret on both the 1997 and '98 expeditions, my primary responsibility was to document the expedition using both digital videography and still photographs. I, along with Marcelo Mendez, both

Fellows in the Explorer's Club and Trustees of the sponsoring museum, filmed the expedition's activities on land and under the sea and in 1998, in the air.

On the Expeditions of '97, we were able to get a pretty fair view of Easter Island from the highest volcanic craters, Rano Raraku, Rano Kau and Rano Aroi. It fired the imagination as to what could be seen if one could fly low and slowly over the Island, map it, and perhaps discover new potential archeological sites. Aerial videography of the many monuments (Moais and Ahus), ancient religious settlements such as Orongo, and many other landmarks, as well as the coastal waters, would add immeasurably to the documentary we were preparing on these expeditions. High definition digital videography offered literally mega-thousands of individual stills for scientific analysis.

The major problem is that Easter Island has no resident aircraft to lease or charter though it does have an airport with an 11,000 ft. runway built by NASA as an emergency strip for the Space Shuttle. Mataveri airport principally serves Lan Chile Airlines, which runs six flights per week. Outside of occasional government planes flying in from Chile, that's pretty much it. There is no reason to keep small planes on the island. It's only 63 square miles in area and the nearest airfield is roughly 2500 miles away with Santiago, Chile to the east, and Tahiti to the west.

For the 1998 Expedition, if we wanted to film from the air, we would have to bring in our own plane by airfreight. The plane would have to be light in weight and easily broken down to fit into a container measuring no more than 4x5x8 ft. It would have to be easy to maintain and considering the two weeks we had to work in, spare parts would have to travel with the plane. We also had to consider the aircraft cost (lease, buy or gifted) and the costs of shipping. The logistics and budget left no alternative other than an ultralight aircraft (FAR Regulation 103 – weight not to exceed 253 lbs.). Considering the volcanic terrain and potential for variable and high winds we ruled out an ultra-light fixed winged aircraft. Our choice was a two place powered parachute with room for a pilot and photographer.

The only ultralight plane with similar characteristics to the parachute plane was the single seat Benson Gyroplane brought into Easter Island by Philip Cousteau in 1972, the son of the famous oceanographer Jacques Cousteau. Unfortunately after a few hours of flight time the plane had an in-air malfunction and crashed. Cousteau, though badly injured, survived to fly another day.

3. PLANE AND PILOT

Good fortune smiled on us a good year before we were to leave for the Easter Island '98 Expedition. I was flying my plane into Bennington Airport, Vermont. I saw a powered parachute about to land on a grass side strip next to the runway. I had flown a similar aircraft in 1984 and I wanted to get a closer look at it. After we both landed I walked over to meet pilot and plane and I introduced myself to Gary Warren who not only flew, but also built and gave instruction on powered parachutes manufactured in kit form by Six Chuter Company, Yakima, Washington.

The Parachute Plane fit all of the qualifications for the expedition. The possibility of bringing a plane to Easter Island was born. As a bonus, Gary was persuaded to join the expedition as pilot and chief maintenance officer.

3.1 The Plane

Letters and phone calls were sent through Gary Warren to the President of Six Chuter, Inc. We explained the purpose and the conditions the plane would be used for in the Easter Island Expedition. We received not only good technical cooperation from Six Chuter, but they agreed to give us a plane for the expedition and solicit additional contributors for the plane, primarily the engine, parachute, and other basic parts to complete the aircraft.

It was all shipped to Bennington, VT, where Gary Warren assembled it. It was test flown and we photographed mountain scenes in Vermont as a further test and then Marcello Mendez, Gary and I, along with the help of local friends, disassembled the plane and carefully packed it in a box measuring 8x5x3 ft. It was then trucked to Lan Chile Airlines, JFK, as freight, to join us on Easter Island.

A portion of the grants raised from The Hale Mathews, Robert Goulet and Jackie Quillen Foundations provided some of the funding for expenses of freight and other plane related expenses. Other costs were personally absorbed.

3.2 Specifications of the Six Chuter SR-5

Visually, a powered parachute looks like a large go-cart with an engine at the rear driving a large propeller housed in a protective tubular ring. The "wing", which is a "flat" parachute, is connected to the plane by a series of risers and is placed on the ground behind the plane prior to take off. The machine itself, and all its component parts are really much more high tech than described above: Some of the more important specs are listed below:

Airspeed	26-30mph
Airframe weight	265lbs
Gross weight	610lbs
Take off distance	100-300ft
Landing distance	50-70ft
Max. Payload	475lbs
Rate of climb	400-700ft
Glide ratio	6:1
Engine (Hirth)	52hp
Parachute span	39.5ft
Parachute area	520sq.ft
Gasoline capacity	9 gallons
Endurance in flight	2.5 hours at max gross

3.3 Other Interesting Features

The plane is steered in flight by "rudder bars" using the pilot's feet. Pushing a rudder bar pulls down a riser attached to the corner of the chute which spills air and turns the plane in that direction – so, right foot turns the plane right, and left foot left. On turns, the plane swings out like a pendulum, creating horizontal lift, similar to a fixed wing aircraft. Altitude is increased or decreased by applying power, but the speeds never deviate much from its top speed of 26 mph. Unlike fixed wing aircraft, powered parachutes are stall resistant. It is a very safe plane to fly even in high winds, although winds in excess of 26mph would mean you would be flying backwards. We added brakes to be able to drive on roads (with the chute stored in a bag). Otherwise we trucked the parachute-plane to a suitable take off strip.

3.4 A Photographic Platform

The Six Chuter SR-5 was test flown many times in the Bennington, VT area. We tried a number of modifications in order to get the best air to ground shots. A clamp-on bracket was tried for a 35mm camera so that slides could be used for documentation. It was set up so that the pilot could trigger the camera with the photographer straddling the pilot from behind and therefore free to concentrate on videography. We also considered using a video monitor attached to the video camera but the tight space and movement from side to side made this impractical.

4. THE CAMERAS

In the selection of our principal camera, we were looking for adaptability and high image quality. It had to handle well in the tight confines of the Para plane. It also had to backpack well for land use, particularly when carried over the rugged terrain of Easter Island's volcanic craters. We found it all in the Canon XL-1, a Digital Video Camera which was light with excellent quality.

This camera combine with the K-6 gyroscopic allowed us to shoot aerial shots in fairly turbulent situations. The results were remarkably smooth. The XL-1 manual control options are excellent; even allowing override while shooting in several automatic modes, often the best of both worlds. Canon's internal optical stabilization system is also remarkable. It deals with the low frequency movements (breathing, bouncing, swaying and even heart beat), and since the system is digital, it does not degrade the image. The XL-1 combines a lot of "professional" features with IC lens capability and superior low-light performance.

Another video camera used in the air and on land was the Sony VX-1 which offered similar quality to the Canon XL-1 and had the added option of time-lapse which we used with great effect on sunsets behind Moai statue groupings. For use in tighter quarters, such as inside caves and lava tunnels, we used smaller 1 chip CCD Panasonic and JVC cameras. Although not as crisp on long shots, they gave us top quality on medium to close-ups. Apart from the 4 video cameras we had the usual accessory equipment: Tripods, sound recording with remote mikes, monitors and field editing equipment.

Rounding out the photographic gear were the still film cameras, 2 Nikon 35 mms (the N90s and the 2000) with the full complement of wide angle and telephoto lenses, filters, and plenty of film.

The flying characteristics of the plane itself posed problems for the photographer, which had to be overcome by pilotage and patience. The body of the plane containing the pilot and photographer is about 18 ft from the overhead chute. Turns and turbulence tend to create extreme roll and pitch, and sometimes a combination of both at the same time. The K-6 gyroscopic stabilizer attached to the video camera, helped smooth out the shots.

It was also important to develop close communication between pilot and photographer. We used headphones with an intercom built into the helmets, which also gave us protection from engine noise. We also could talk to the control tower and our ground mobile station – a 1994 Geo Tracker. One of our hand held Bendix King KX99 Nav-Com radios was tied to the plane's airframe. The other radio was kept in the Tracker. One of the three of us always drove the truck, mostly off road, as we followed the plane. The truck carried gasoline, spare parts, tools, and a first aid kit – plus lunch and drinks (soft). Following the truck, which was following the plane, was a host of Rapa Nui natives in cars, horses, or on foot. Flying low and slow and somewhat noisy, the Paraplane always drew a crowd.

Communication between plane and control tower gave us weather updates while we were in the air and was a safety link in case of an accident. It also gave us take-off and landing clearances. Considering the tower only handled 6 to 7 flights a week, you can imagine how much they enjoyed the contact with us. Even a medium sized airport such as my home base in Danbury, Connecticut would often handle 6 to 7 different flight contacts in a minute. At first we exchanged a formal exchange such as "Mataveri Tower, this is Paraplane 1 ready for take off". The Tower would answer in international aviation language – English – something like, "Paraplane 1 cleared for take off on runway 27, altimeter 29.2 – winds 230 (direction) at 10 (knots)". After we spent some personal time with the tower personnel, we became less formal with our language; after all we might be their only customers for the day.

5. GROUND TARGET OBJECTIVES

Outside of the broad coverage of the island, our principle targets were: the major Moia concentrations including standing statues, ruins and Moai in situ, and the immediate surrounding landscape; the coastline, including tidal pools, principle beaches and small boat ports; prominent land targets with special emphasis on the three major volcanic craters along with the Poike area, Hanga Roa, Puna Pau, and a decent representation of the interior island landscape.

We managed to meet our target objectives within the limited time of some 10 days out of the 15 we were there. We were fortunate to have overall good weather. We did hit a few windy days when we couldn't fly, and other days where flying was limited to early morning and/or early evening. Also, it was not always sunny, and that does make a difference in aerial photography. Flat light does not bring out the three-dimensional effects you are looking for. Neither does midday sun. Overall we shot about 8 hours of aerial footage out of 20 flying hours, and have some exclusive low level video and photographs, which will be used in a documentary we are preparing. We had the opportunity to supply a CNN production crew, who was covering our expedition, with land, underwater and aerial footage. They used it in program entitled, *Forging Ahead* which aired worldwide on November 8, 1998. We also supplied footage to CPG Productions for a History Channel documentary entitled *The Mysteries of Easter Island.*

6. CONCLUSIONS

There is no doubt that the two man powered parachute we took with us to Easter Island was perfect for our tape and filming objectives considering budget constraints. It was low cost to pack, freight, and maintain. The Six Chuter parachute plane was strong and safe. When faced with high winds we were able to put it down anywhere there was a flat surface with minimal space. When we did fly over water to photograph the coastline, or the islands off Orongo, we made sure the winds were blowing inland. A parachute plane floats about as well as a motorcycle. The waves are high and the coastline is battered by heavy waves. If we lost power we wanted to glide inland, and after all, we were in a parachute. We flew in most all kinds of weather, including rain, with no serious accidents. We also had an excellent pilot and maintenance man in Gary Warren.

Outside of our use of the plane for photographic purposes, we were able to use the plane as an observation post, and gave a number of Rapa Nuis a chance to see their island from a low and slow moving platform. We also gave some of the expedition members a chance to fly. This was particularly valuable for Dr. John Loret, co-leader of the expedition, and the members of the U.S. National Park Service headed by Kevin Buckley, who was there to work with the local park management and their Chilean counterpart, CONEF. We also drove the plane through the town of Hanga Roa where we addressed a group of students who were members of a newly formed Air ROTC unit. Our biggest concern was that the plane, because it flew low and slow (from 300 to 2,000ft), would be an irritation to the Rapa Nuis. This was not the case. To many on the island we became The Birdmen – The Tangata Manu. Everyone from the Governor to the military and airport management and traffic controllers couldn't have been more helpful. They supplied us with a hangar on the airport grounds, and much needed weather information.

Not the least of this venture were the many new friends we made with the native population and the opportunity to share our adventure with them both while we were there and in the pictures and tapes which were sent back as a shared memory our visit.

Chapter 12

Easter Island Under Glass: Observations and Conversations

LINDLEY KIRKSEY
Explorer's Club, New York, USA

1. INTRODUCTION

Mysterious Easter Island? How the people will embrace the 21st Century is the real mystery. Electricity arrived in the 70's, taxis and the Concord in the 90's. And the Internet arrives in a few weeks. The beauty of the flowering African Tulip tree and the serenity of a lone horseman crossing the ceremonial field at Tahai at sunset remain as first impressions. One image represents the impact of introduced culture to the island; the other, the communion of man with his land. Both are key to an understanding of Easter Island today.

Easter Island, a province of Chile and the most remote civilized island in the world, is positioned approximately 2200 miles west of Chile and 1400 miles east of Pitcairn Island. 42% of Easter Island is national park and a World Heritage Site with most of the population living in the town of Hangaroa, where island services are concentrated. The rise and collapse of the native civilization between about 400 AD and the first European contact in 1722 present unanswered questions: the Polynesian origin of the first settlers, the carving and movement of stone statues weighing up to 80 tons, or more, the decline in the culture, and the depletion of natural resources. All have contributed to the mystery surrounding Rapa Nui.

Easter Island, Edited by John Loret and John T. Tanacredi
Kluwer Academic/Plenum Publishers, New York, 2003

Figure 1. View of Tahai, showing two of the *ahu* at this complex. Close to the village, the site receives a lot of visitation by tourists.

As part of an interdisciplinary team sponsored by the Science Museum of Long Island and led by John Loret and Robert Hemm, I participated in a scientific update of the events that led to the precontact drop in population and over use of natural resources. Today, no native trees exist on this once wooded volcanic island. A native population that once numbered over 10,000 people now numbers about 3000. At its lowest point, due to disease, lack of food, civil war, and slave raids, the population had fallen to a mere 111. Such a rise, fall, and recent recovery in a society isolated from outside contact for over 1300 years suggests a model for the progress on Planet Earth – of man's interaction with his resources. In pondering this, I realized what a rare opportunity I had to roam this peaceful place and experience the openness and warmth of the Rapanui. The themes that reoccurred in conversations – spontaneous and arranged – about the island were land ownership and preserving a sense of personal freedom as the island people struggle for identity and survival.

How do the Rapa Nui experience themselves in the world? An early name for the island was Te Pito 'o te Henua, or Navel of the Earth. They feel themselves to be a special people and cherish the land they belong to. This is evidenced in how often they return. Those who move to Tahiti or the Continent immediately plan a return visit and many would prefer to die where they were born. The present is more real than the future and myth and reality sometimes are interwoven. The island priest, Father Joe Navarrete, and H. Chivano, teacher of philosophy at the high school, both spoke of the importance of dreams and of the spirit world where good and bad spirits are given deference. Even today an elderly person may simply wander into the interior of the island to find a place to die – a place of clan significance or a secret cave. Once the entire school turned out to look for a woman who was never found.

2. THE RAPA NUI PEOPLE

Friendships and business relationships depend on certain underlying structures affecting daily life, as they do with all people, and these need to be understood. Family life on the island can be a fluid arrangement. Many couples do not marry even though they may share a house and raise children. Most marriages today are mixed with one spouse being from off the island. This is understandable when one considers that almost all the people on the island are, by now, related and that cousins who marry can be ostracized by the rest of the community. Several women that I spoke with did not want to marry a Rapa Nui man because there can be domestic abuse between husband and wife. I heard this from many people, including the Catholic priest. Tribal custom dictated that the woman became the property of the man – a loss of freedom. A welfare system is not needed on the island as the tribal system is also a support system. Women often follow the Chilean custom of keeping their maiden name. This also serves to preserve clan recognition. For a child, all aunts and uncles are parents and all adults are aunts and uncles. A child may go from home to home for any number of

Figure 2. A group of preschool children on an outing in Hanga Roa's small park.

reasons, residing there as a member of the family – fed and cared for. Accustomed to having the run of the whole island, school children – who live in constant contact with the environment – pose disciplinary problem for the teachers and are defended in their freedom to roam by parents and guardians.

Discipline issues aside, parents are genuinely interested in the education of their children. Eight hundred students attend grades kindergarten through grade twelve. Most finish high school and some receive scholarships to go to college in Chile. The Chilean government is encouraging unique, regional approaches to curriculum development, a concept that may well preserve the cultural heritage of the Rapa Nui. Parents and staff are meeting now to address these critical areas: preparation for university, personal relations, environment, preservation of the island, awareness of old people, respect, and preservation of the language. Language in general is a major concern. In the home, both Spanish and Rapa Nui are often spoken. Recently, English has been required through grade twelve and French is an elective. Many islanders have ties in Tahiti, and at least French and some Japanese language skills are picked by those engaged in the active tourist business.

According to Father Joe Navarrete, the Islanders are good church attendees – for 100 years most islanders have been Roman Catholic – though many do not observe the sacrament of marriage. I observed children learning to sing the hymns, most of which are in Tahitian, and attended the family service. The Mass is in Spanish and some Rapa Nui. Ten songs are in Rapa Nui. This mixture of language and culture extends to the carved

Figure 3. Rapa Nui children in the island's Catholic Church. The elegant fish on the wood lecture was carved by local artists.

wooden statues that adorn the church. Carved in the island tradition, they contain symbols from the ancient cult religion of the Birdman, a cult image that emerged at a time when civil war and hunger were prevalent. The warm, mellow tones of the wood are in contrast to traditional Spanish-Catholic religious statuary.

There are no homeless on Easter Island and no begging children. To compensate, the informal system is based on a sharing of items and services. A colorful example is this: A man may request of another a pack of cigarettes – the entire pack, not just one cigarette. If the owner has a pack, he will no doubt give it over. A few weeks later, the giver may be helped with a tire repair on the road by the receiver of the pack of cigarettes. This concept of asking is complex and pervasive. All manner of jobs and favoritism go first to one's kin. Ownership of land is confined to those of Rapa Nui decent and no taxes are collected. The island is administered as a province of Chile with an appointed governor and an elected mayor and council members. The islanders therefore feel both protected and provided for and at the same time prevented from the free use of their own land and from the planning of their own destiny. Fishing and tourism are the main income sources for the island and since approximately 90 babies are born each year, I asked then Mayor Pedro Edmunds Paoa if he felt that land would become scarce. With a smile, he informed me that Easter Island is the size of Hong Kong and that land parcels could be given out for 200 years. In the middle of the Pacific Ocean as it is, it is not likely that Easter Island will become the hub that Hong Kong is, but I got the point.

3. RAPA NUI NATIONAL PARK

Land really is the issue on Easter Island. With over 10,000 tourists arriving by plane or ship each year and with almost half the land deemed national park, the management of this fragile resource is at the heart of the economy and future of Easter Island. The Corporation National Forestal, CONAF, is the department of the central government of Chile that administers the Rapa Nui National Park. Approximately 20 people are on the staff on Easter Island. Marcos Rauch Gonzalez, the archaeologist and second in command spoke candidly about the need for training in conservation of sites and in interpretation of both the archaeology and the environmental issues. Over 800 statues (*moai*) mostly carved from volcanic tuff, and 300 ceremonial platforms (*ahu*) plus over 5000 petroglyphs, quarries, volcanic caves, and simple or informal sites are under the protection of the park system. Livestock, tourists, and local people roam freely about the island leaving sites in the middle of fields especially vulnerable. Animals are particularly destructive as are erosion and fire. In 1996, a fire in the volcanic crater Rano Raraku scarred one *moai* and affected 46 others. The lone park ranger could not call for help since CONAF personnel on the island do not have radio transmitters.

Figure 4. A Makemake face painted by local islanders for the benefit of the tourist cameras.

Three locations have on-site rangers: Orongo, a ceremonial center connected with the Birdman cult; Rano Raraku, the quarry where the massive status were carved in one piece from volcanic tuff; and Anakena, the beach where Hotu Matua, legendary founder of the Rapa Nui people, landed. Even so, theft and vandalism are hard to prevent since there is no authority for protection and the role of ranger is not that of a policeman. At 'Orongo, the interpretation plaques were stolen and a rare artifact from the Easter Island Museum was taken by a young boy who smashed it in his escape. Knowing that visitors like "discovery", animal bones are sometimes deposited in niches at the ceremonial sites where the dead were once buried. There is an interesting forerunner to this activity. During Thor Heyerdahl's expedition in 1955, islanders would work all night to produce the sought "secret cave" artifacts, appropriately distressed.

This is not a case of the outside world not being interested. Easter Island is one of the most studied places on earth. UNESCO and the World Monuments Fund have provided funds for studies and restoration projects and recognize that heritage and preservation of the island's treasures are international concerns. Sergio Rapu, archaeologist, local businessman, and former Governor of the island, would like to see park people trained in the American National Park system through funds sent directly to the island. Too often up to 50% of funds designated for Easter Island remain in Chile for administrative purposes. Because the municipal law permits the city to form an outside non-profit corporation, the Mayor and Sergio Rapu both support setting up a foundation funded by private enterprises to be managed for the benefit of Easter Island and the National Park System. A council would determine where and how the income would be spent. Preservation is not a casual undertaking. To restore a monument at Anakena it took $400,000. Fourteen years ago, $16,000 was spent to preserve one statue. This activity is on hold since the consequences of preservation are not known, and because the cost of treatment is high.

Something should be mentioned about stone conservation – a controversial issue. Most of the large *moai* on Easter Island are carved from volcanic tuff and are showing serious deterioration. Though spared the ravages of freezing weather, the winds, rain and salt water are constant and take a toll. Applying coating to stone can form a barrier to the elements and protect stone for a while. Various alkoxysilanes are often the choice and consolidate the stone by forming a bonding with the surface. The idea is to keep out moisture and preserve the surface. A successful conservation project was carried out on one *moai* at Hanga Kio'e. Further restoration has been deterred because of unanswered questions: What happens ten years from now? Can the process be undone? Will it wear off unevenly and leave a blotched effect? Part of the process of time is that all things age. An alternative to massive (literally) restoration projects would be to enclose some statues, make models of certain statues, and record accurate information on film and in writing using sophisticated equipment. Objects can be scanned and recorded in digital form. Excellent material on this topic can be found in the publications of The Getty Conservation Institute and those of the World Monuments Fund, specifically, *Easter Island: The Heritage and its Conservation* by Dr. Elena Charola.

4. EASTER ISLAND ECONOMY

In considering the economics of life on Easter Island in the next century, it is important to see the need to further the development of tourism and also to expand income producing activities for the islanders who hunger as much as the rest of the world for refrigerators and TV's. The soil is thin, erosion a problem, and fresh water is found only underground, in the volcanic craters or catchment water holes. With the frequent airplane deliveries, islanders often spend more on imported foods than on the local produce. To address these issues, Sergio Rapu suggests the founding of a technical institute on the island to train young people in the history and archaeology of the island and in private enterprise. Again, this is seen as a road to freedom. Since 1966, islanders can vote and travel, but they cannot sell their own land or go to the bank and take out an equity loan. Sergio would want to introduce the

Figure 5. Sergio Rapu (center) discussing plans to restore the *ahu* at Hanga Piko. On the right are Dr. John Loret and Mayor Pedro Edmunds.

technology of growing hydroponic vegetables, and an authority on this process has already visited the island from Israel. Sergio would also like to train young people to be "archaeological technicians" – able to speak with authority about the island's treasures.

No one on Easter Island is rich, and those who are well off are in the tourist business. To boost economic development the Mayor of Easter Island is encouraging the best of the island's carvers to produce the elegant walking canes that include the ancient motif and is undertaking a tour to secure upscale markets for these products in the United States. Because visitors seldom stay on the island more than three days, the Mayor also wants the island to offer walks to the less accessible sites, trail rides for horseback riders, and nature and bird watching outings. Certain monument sites, with the sea beyond, lend themselves as outstanding outdoor performance areas for use by the local people as well as the tourists. I would hope that a park plan would include a welcome center in Hangaroa where accurate and specific information and videos, all in a number of foreign languages, would orient visitors and equip them for outings on this unique island.

Another economic issue for Rapa Nui is the lack of a deep water port. Tourists arriving by ship must be ferried from the ship to the island, and lighters are used to bring in goods from container ships arriving from the continent about three times a year. It costs as much to have the lighter bring the containers from the ship as it does to bring the ship 2200 miles from the continent. Plans are to build a new deep water port at La Perouse Bay, on the northwest coast. An article in *Rapa Nui News* (see the Easter Island web page on the Internet) reports that the proposal has the island divided between those wanting "progress" and those who want to preserve the pastoral beauty of the natural area – and one that the tourists come to see. Some feel that the already existing port at Hanga Piko should be expanded to accommodate larger ships.

This division on the issue of a new port is but one example of internal strife that can flare up at any time over large and small issues. A major division at the moment concerns the Council of Elders. A second Council was formed in opposition to the policies of the first, so there are Councils #1 and #2. One might think that the old tribal warfare has broken out again, rupturing families and pitting cousin against cousin. It would seem that hot tempers and infighting among the islanders have to be ameliorated before the Chilean Government eases its control of the island's affairs.

5. THE FUTURE

Another area of particular concern to me was the lack of knowledge young adults on the island have of historical events and of volcanism. When asked about some aspect of the island, the secret caves, for instance, they are apt to give an answer from their ancient mythology and no seeming awareness of how the caves were formed or that they were places of refuge when slave hunters visited the island and in times of civil strife. In his book, *Rapa Nui: Tradition and Survival on Easter Island*, anthropologist Grant McCall mentions being asked by a highly educated Rapa Nui man if the Peruvian slavery really took place. This poses an interesting question. In this age of revisionist history, is it possible to give the Rapa Nui, who experienced extreme exploitation, a history they can be proud of? Perhaps this can be eased by defining acts of courage and by emphasizing the positive characteristics that continue to sustain the Rapa Nui who have survived.

Fortunately for the rest of the world, the Rapa Nui have a strong identity that they fight to preserve. This is in concert with their natural desire to acquire the benefits of the modern world – no small undertaking in a place where a pair of blue jeans cost $70 US. With off-island marriages and fewer children speaking Rapa Nui at home, it is reassuring to witness their desire to preserve the language and history (as seen in the Tapati Festival). Some of this motivation may be economically driven as the islanders know well the lure the mysterious island has for the tourist. International organizations also recognize the need to help preserve this remarkable heritage.

Change will come to the island. It is desired, necessary, and inevitable. I learned that 150 endemic *toromiro* trees, reintroduced to the island by well-intentioned botanists in Bonn, had all died. I could not resist the idea of Easter Island as the world's terrarium. In the effort of preservation, care must be taken to help – but not to impose – while leaving the destiny of the Rapa Nui in their own hands.

6. UPDATE (YEAR 2000)

Father Francisco Nahoe and Grant McCall report that the number of Rapa Nui now living on Easter Island is approximately 3000 and growing. An additional 300 Rapa Nui live in French Polynesia and approximately 700 live in Continental Chile. They located one Rapa Nui who has lived in Israel for 25 years.

Figure 6. Tourists at Rano Raraku's statue quarry climb and walk on the statues that still lie in the rock; there are no barriers and little supervision.

According to the *Rapa Nui News, Number 18. March 30, 2000,* 1,500 hectares of land have been restored to the Rapa Nui community fulfilling a promise made nine years ago by former Chilean President Aylwin. This represented five hectare-sized farms that were distributed among some 280 previously selected islanders who participated in a lottery headed by the Provincial Governor, Jacobo Hey Paoa. This is a beginning, as there still remains a great yearning among the islanders to be proprietors of their own land. The law still forbids the selling of land to foreigners, including Continental Chileans.

In a move to protect archaeological sites close to the roads from vehicles, stone walls have been erected at sites including Vaihu, Akahanga, and Te Pito Kura.

Discussion continues on the island to establish telemedicine following the death of a patient who was transferred to the continent. Telemedicine would enable a doctor at the Hanga Roa hospital to make audiovisual contact with specialists off the island when there are complicated cases to be resolved and diagnostics are needed.

Photographs by Lindley Kirksey and Holly Williams.

REFERENCES

Charola, A. Elena, *Easter Island: The Heritage and its Conservation.* World Monuments Fund: New York, (1994).
Fisher, Steven Roger, ed, *Easter Island Studies.* Oxbow Books: Oxford, UK, (1993).
McCall, Grant, *Rapa Nui: Tradition & Survival on Easter Island.* University of Hawaii Press: Honolulu, (1994).
Price, C.A., *Stone Conservation: An Overview of Current Research.* The Getty Conservation Institute: Santa Monica, CA, (1996).
Van Tilburg, Jo Anne, *Easter Island: Archaeology, Ecology and Culture*, Smithsonian Institution Press, Washington, DC (1994).

Chapter 13

Mapping The Poike Ditch

RICHARD E. REANIER[1] and DONALD P. RYAN[2]
[1]*Reanier & Associates, Inc., Seattle, Washington, USA;* [2]*Pacific Lutheran University, Tacoma, Washington, USA*

1. INTRODUCTION

Rapa Nui (Easter Island) is known world wide for its huge stone statues, the *moai*, which grace the covers of numerous books and lure travellers to the island. These monuments are but one of many kinds of archaeological features of the island. Others, such as the *ahu* platforms upon which the ancient *moai* rested, the beautifully restored stone house complexes at Orongo, and a wealth of petroglyphs are well familiar to visitors to the island. These remains, and many others, reflect a remarkable record of more than 1,500 years of human history on this remote dot of land in the southeast Pacific.

Yet it is another, relatively unimposing feature that is the subject of this paper – the Poike Ditch. Today, this feature is a discontinuous linear string of 20 or so elongated depressions that stretch for nearly 2 km across the base of the Poike peninsula between Hotu Iti on the south and Mahatua on the north (Figs 1 and 2). Though relatively obscure when compared to other archaeological features of the island, the Poike Ditch plays an important role in Rapa Nui oral history, and also has provided archaeologists with the earliest radiocarbon date for human presence on the island.

Easter Island, Edited by John Loret and John T. Tanacredi
Kluwer Academic/Plenum Publishers, New York, 2003

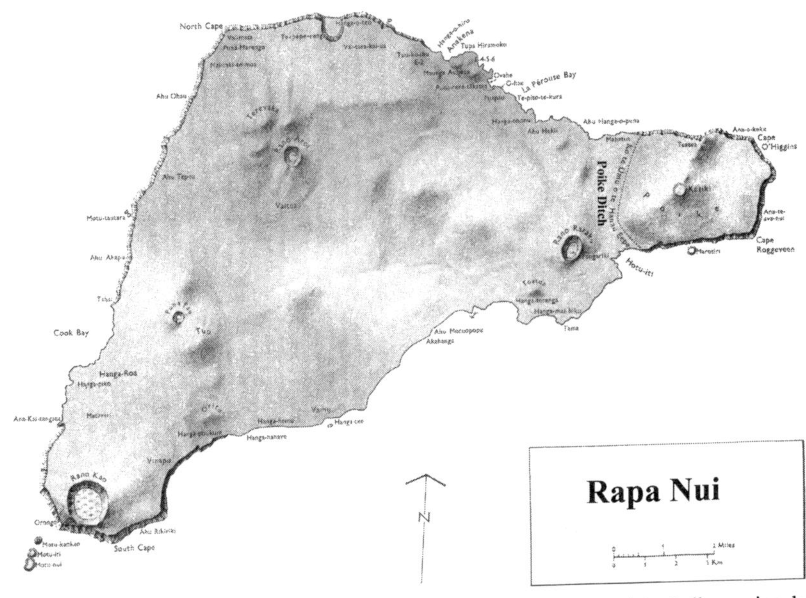

Figure 1. Map of Easter Island. The Poike Ditch is located at the base of the Poike peninsula on the eastern end of the island (after Heyerdahl and Ferdon, 1961:22).

Figure 2. The Poike Ditch as it appears today, with the crater Rano Raraku in the background. The berm or spoil pile is upslope on the left, and the trench, here not very pronounced, is on the right.

There have been a few limited investigations of the feature, beginning with the Routledge expedition to the island in 1914-1915 (Routledge 1919:281). This expedition examined the ditch and may have conducted some test excavations (Smith, 1961). During the Norwegian Archaeological Expedition led by Thor Heyerdahl in 1955-1956, archaeologist Carlyle Smith excavated six test pits in the ditch and a 30 m long trench across it in an effort to understand its origins. A team from the Universidad de Chile also excavated two trenches across the feature within the last two decades.

In August of 1998 we decided that creating a detailed map of the Poike Ditch would be a valuable contribution to the archaeology of Easter Island. Although the feature had been excavated at least twice, there was no published map of its entire length – and no detailed topographic map of the feature at all. The reason for our concern was simple. Most of the other archaeological monuments on the island are made of stone, but the Poike Ditch is only an earthen feature, and therefore is subject to destruction from erosion. In 1955, Smith noted several periods of deposition in the ditch resulting from erosion, the most recent of which was likely due to overgrazing by sheep in the 20th century (Smith, 1961). He estimated 85 cm of soil had been deposited within the ditch in the past 280 years. Today, the effects of erosion are clear. Sedimentation has continued to occur within the swales, and erosion is also at work in places, removing soil from the berm or embankment that forms the upslope side of the feature. By creating a detailed topographic map of the feature we could digitally preserve the morphology of the ditch as it is today, whatever its erosional fate in the future.

2. KO TE UMU O TE HANAU EEPE

The Poike Ditch is known in the Rapa Nui language as *Ko te Umu o te Hanau-Eepe* or "The Earth Oven of the Long-Ears." This name refers to an important story in Rapa Nui oral history, the story of a war between two groups of people: the Long-ears (*Hanau eepe*) and the Short-ears (*Hanau momoko*). There are several published versions of this story, the earliest of which was recorded by Thompson (1889) who visited the island briefly in 1886. Routledge, who spent 18 months on the island in 1914-1915, published a brief account of the story in her book recounting the journey (Routledge, 1919), and other versions have been published by Knoche (1925) and Vives Solar (1930). Métraux provided an analysis of these versions of the story in his monograph on the ethnology of the island based upon fieldwork he conducted there in 1934-1935, and compared them to the version he recorded (Métraux, 1971).

According to the story, the ditch was a defensive structure constructed by the *Hanau eepe*. The trench, it was said, was filled with flammable material that could be ignited in a conflict with the *Hanau momoko*. During such a battle, the trench was ignited and the *Hanau eepe* were forced into their own burning ditch and only a couple of their kind survived. The validity of the legend has been disputed and the true nature of the ditch remains controversial.

3. POIKE DITCH ORIGINS

Based on her study of the feature and her knowledge of the oral tradition, Katherine Routledge (1919) concluded that the ditch was a geologic feature caused by faulting, but that it might also have been used as a defensive feature by people protecting the eastern slopes of the Poike peninsula. One of the principal features of the ditch is that the soil removed by digging (the spoil pile) has been placed on the upslope side of the ditch. This would have taken considerably more effort than piling the soil on the downslope side, since one must lift each load of soil higher to place it on the upslope side than on the downslope. It seems reasonable that a ridge placed along the upslope side of the ditch would serve as a more formidable barrier than one placed on the downslope side. Although in Routledge's version of the Long-ears and Short-ears story the Long-ears perished by being burned in the ditch, it does not mention the ditch being used as an earth oven. Twenty years later Métraux's informant specifically indicated its use as an earth oven, as the title given to the story shows.

The geologist Lawrence Chubb, who visited Rapa Nui in 1925, described the Poike Ditch as a gully running from north to south along the base of the Poike peninsula that was formed by recent lava flows from Terevaka volcano on the west and the ancient wave-cut cliffs of Poike on the east (Chubb, 1933). He described exposures of the ancient wave-cut cliffs of Poike as extending along its entire length (Chubb, 1933). However, there are no exposed wave-cut cliffs anywhere along the length of the ditch, suggesting that Chubb did not actually examine the Poike Ditch during his study.

Carlyle Smith with the Norwegian expedition completed the first trench across the ditch in 1955. His cross sections revealed the elongated mounds on the upslope side of the ditch were anthropogenic, and that the ditch had been dug to a depth of more than 2.5 meters below the surface of the 1955 depression (Fig. 3). This 2.5 m of fill contained artifacts and layers of charcoal (Smith, 1961). Smith's work also showed that the ditch likely has a natural origin. Below the depth to which the ditch had been excavated in prehistoric times Smith found evidence that soil devoid of artifacts had collapsed into a large crack between two faces of lava, filling a gap more than 5 m wide (Fig. 3). Thus, Smith's evidence confirmed Routledge's hypothesis that the ditch is a natural feature later modified by the island's inhabitants. Smith concluded that the ditch was probably used as a defensive fortification (Smith, 1961), but reflecting on the work some 30 years later he revised his ideas on the matter. In 1990 he considered its use as a fortification less likely than its use as a moist swale for planting crops to feed the workers carving *moai* at Rano Raraku, or as an earth oven for cooking such food (Smith, 1990). Whatever the origins of the Poike Ditch, it is useful to note that these hypotheses, and others, are not mutually exclusive. In the 1600 years of human presence on the island the feature, easily recognizable to the island's inhabitants, could have been put to many uses.

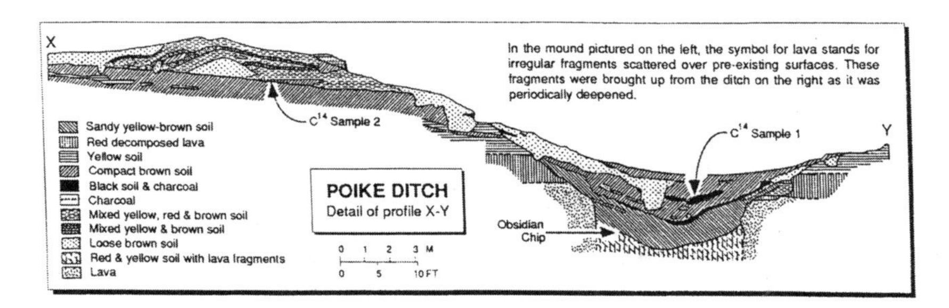

Figure 3. Stratigraphic profile of Smith's trench across the Poike Ditch. Sample 1 from within the ditch fill dates to 280 ± 100 B.P., while Sample 2 from beneath the spoil pile dates to 1570 ± 100 B.P. (From Smith, 1990:31, courtesy of Rapa Nui Journal).

Smith's work also resulted in two radiocarbon dates for the Poike Ditch. One charcoal sample from the ditch fill, between 85 cm and 100 cm below the surface, produced an age of 280 ± 100 B.P. (K-501) [1] (Sample 1, Fig. 3) (Smith, 1961). This sample consisted of "burned twigs, stalks of cane like material, leaves, and grass in a compact mass interspersed with lenses of earth turned red from contact with intense heat" (Smith, 1961). This date seems to be in agreement with Father Sebastian Englert's estimation by means of genealogical data that the Poike battle would have taken place in the latter part of the 17th century (Englert, 1970). The other charcoal sample submitted for dating came from the original ground surface over which the

[1] B.P. is a radiocarbon dating convention that signifies years before present, where the present is taken to be A.D. 1950. The number in parentheses, in this case K-501, indicates the laboratory number of the dating determination.

initial spoil from the ditch had been dumped at the beginning of the prehistoric excavation of the ditch (Sample 2, Fig. 3) (Smith, 1961). Smith emphasized that there was no fire reddening of the surrounding soil, indicating the charcoal did not burn *in situ* but was simply lying on the ancient surface of the ground before it was buried by spoil from the ditch. The sample was comprised of large pieces of well-preserved wood charcoal. This date, 1570 ± 100 B.P. (K-502), still stands as the oldest cultural date on the island. These dates reveal the ditch was initially excavated nearly 1600 radiocarbon years ago, and that fires were still being burned in it as recently as 280 radiocarbon years ago.

4. FIELD METHODS

The question of the origins of the Poike Ditch surely remains intriguing and merits further research, but our task in 1998 was to focus on recording the topography of the ditch. Our mapping procedures in the field followed conventional methods for constructing topographic maps. Using a surveying instrument called a Total Station, we recorded the latitude, longitude, and altitude of more than 2,100 points on the surface of the ground along the 2 km length of the ditch. The points were chosen to record the overall shape of the westward sloping terrain, and to capture in greater detail the shape of the ditch itself. A Total Station is a surveying instrument that records angles and distances from the instrument to points on the ground marked by a prism mounted on a pole that reflects light projected from the instrument back to it (Fig. 4). Conveniently, this modern instrument can automatically convert the angles and distances to a Cartesian coordinate system, so that the data are recorded in X, Y, and Z coordinates. The instrument records data quickly, and most of the field time is taken up by the person carrying the prism to the next point to be recorded. Taking this into account, we often had four prisms in use at the same time, one being read while the others were being moved to their next locations. In all, the data recording took eight days, with many volunteers, directed by radio because of the distances involved, maneuvering the prism poles from south to north along the Poike Ditch.

Figure 4. Topcon Total Station being tested at Rano Raraku (top panel). Total Station set up at Maunga Kororau for the 3.6 km shot to tie the map to geographic features of the island (lower panel).

Although collecting data for the topographic map was simple enough in principle, working in any remote location presents its own challenges. Easter Island is no exception. To begin with – we needed somewhere to begin. Conventionally, one begins at a known location – a survey monument – for which the latitude, longitude, and altitude are already known. Proceeding from this known point, each of the subsequently recorded points is therefore recorded in the same, known spatial system. However, on the maps of the Poike area of Easter Island there are no survey monuments marked. During our mapping we encountered several temporary survey monuments, but we were unsuccessful in finding a record of their known locations, if in fact they were known. So, we began our surveying project by using an arbitrary coordinate system – one which is very precise, but which is not connected to "real" latitude, longitude, or altitude. This problem we would fix later.

Using this arbitrary grid system we could produce perfectly good maps – but only for the narrow strip of land we had mapped. We wanted to "connect" our maps to the existing topographic maps of the island, so we could view the Poike Ditch in relation to the surrounding terrain. To do this we connected our arbitrary mapping grid to prominent geographic features that we could also locate on the existing topographic maps. We chose the two closest suitable geographic features marked on the maps: Maunga Kororau, a small hill just west of our mapping area, and Maunga Toa Toa, a prominent volcanic feature southwest of Rano Raraku crater. These two features had the added advantage of having their altitudes already marked on the existing maps. Connecting these features to our mapping grid was something of a technical challenge because of the long distances involved. Our final shot with the Total Station was from Maunga Kororau to Maunga Toa Toa, a distance of more than 3.6 km (Fig. 4). To do this we needed more light reflecting power than a single prism provided, so we attached three prisms to a single pole. Even though the day was windy, shaking both the pole and the Total Station slightly, we were able to get a good reading on the first try: 3,632.61 m.

But, where exactly are Maunga Kororau and Maunga Toa Toa in the real world? To answer this question we might have measured their positions on the topographic map of the island, but this would have been rather imprecise. Instead, we used the Global Positioning System (GPS) to precisely locate the very points we had measured on the two geographic features with the Total Station. Every hiker is now familiar with the small, handheld GPS receivers that pinpoint locations using satellite ranging. Land surveyors and mapmakers use more sophisticated versions of these devices for precise positioning. We used mapping-grade GPS receivers and a method called differential correction to locate Maunga Kororau and Maunga Toa Toa (Fig. 5). Differential correction requires the use of two GPS receivers, one set up

known with very high accuracy, and so we used a survey monument spatially linked to it as the location for our base station (Fig. 6). While our base station was recording GPS data we went to both Maunga Kororau and Maunga Toa Toa and recorded data with the rover station. Later, we used carrier phase processing to determine the exact locations of these two geographic features.

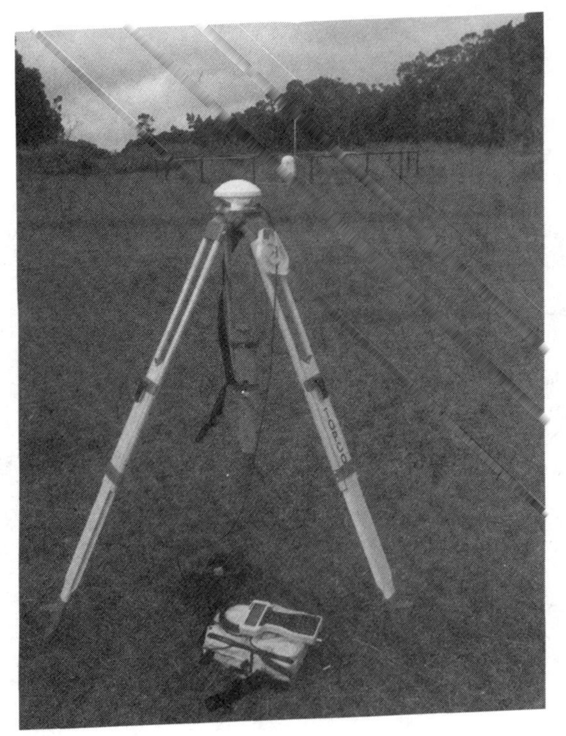

Figure 6. Trimble Pro XL receiver recording base station data at the NASA GPS station. The NASA GPS antenna is visible in the background.

At this point we had a good mapping data set for the Poike Ditch, and we had located two geographic points in this data set with precise GPS positioning. From here it was relatively straightforward to convert our arbitrary mapping grid by shifting and rotating it so it became latitude, longitude, and altitude. The remaining data collection task was to digitize portions of the existing 1:10 000 scale topographic map of the island so that we could view our long and narrow map in the context of the broader geographic features of the Poike area.

5. RESULTS

The final map of the Poike Ditch, more than a meter in length, is too large to reproduce here. However, examination of a portion of the map and the surface model used to produce it will give the reader a clear view of the results. Figure 7 shows the location of our mapping area at the base of the Poike peninsula and areas where the Poike Ditch is particularly visible on the land surface.

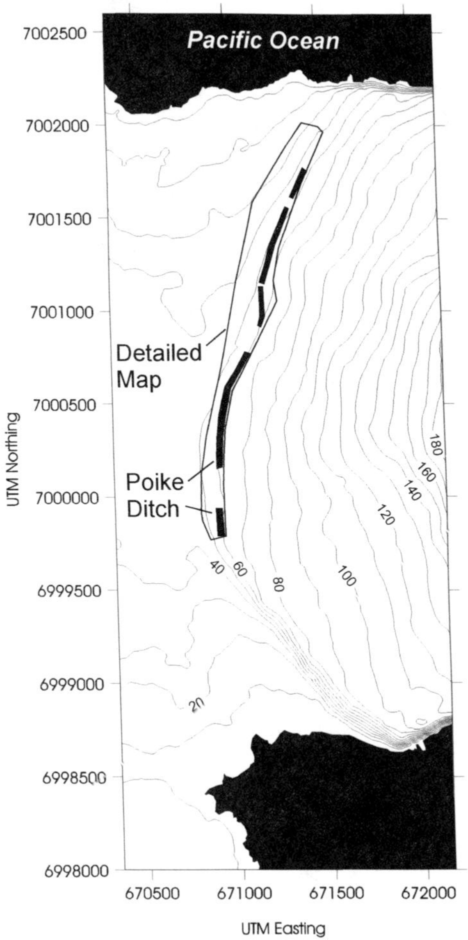

Figure 7. The location of the detailed mapping area at the base of the Poike peninsula. The western slopes of Poike are visible on the right side of the map. Black rectangles indicate areas where the Poike Ditch is particularly conspicuous on the modern land surface. Contour lines are in meters, and geographic coordinates have been converted to the Universal Transverse Mercator (UTM) grid.

Our mapping project also adds to discussions of the origin of the ditch. Figures 7 and 8 indicate the ditch lies not at the base of the slope, but about 20 m above the base of the lower slopes of the peninsula, as Smith had observed earlier (Smith, 1961). From his trench excavation Smith concluded that soil below the cultural layers had filled in a wide crack between two bodies of decomposing volcanic bedrock long before humans utilized the feature. South of the Poike Ditch one can clearly see the southern wave-cut coastal bluffs of Poike trending inland and aligning with the southern end of the ditch itself (Figs 7 and 8). This suggests that the ditch marks the continuation of the ancient western coastline of what would have been the island of Poike before it was joined to the rest of Easter Island by lava from later eruptions of the Terevaka volcano. Today the smooth, soil-covered slopes of Poike contrast markedly with the rugged terrain of the recent lava flows west of the ditch. If the Poike Ditch delineates the ancient western coast of Poike, perhaps the ditch itself was formed by earthquakes accompanying one of the more recent eruptions on Easter Island at a time after the island of Poike had been joined to the main island. Soil from Poike covering the ancient and now-buried coastline might have slumped into a crack opened during seismic activity. This could account for the infilling of

Figure 8. Surface plot of the topographic information in Figure 2 showing the location of prominent sections of the Poike Ditch along the western slopes of Poike. Note the wave-cut coastal bluff south of the ditch.

soil in a bedrock crack beneath the cultural layers that Smith noted during his trench excavations (Fig. 3). Later human modification of a natural system of swales would have resulted in the feature we see today.

Our map results also preclude the use of "ditch" in the traditional sense of the word for transporting water from one end of the feature to the other. Earlier, Smith had concluded the berm on the upslope side of the ditch would tend to prevent water from entering it (Smith, 1961). Figures 7 and 8 show the ditch rises about 5 to 8 m to a high spot located about midway between its north and south ends. This would cause any water entering the ditch to flow north and south from this center point toward the ends (if it were not first absorbed by the soil).

Carlyle Smith's 1955-56 trench across the Poike Ditch was at its southern end (Figs 7 and 8), and we were able to relocate it and several other features along the ditch and record their positions on our map, tying our map to earlier research along the ditch. Figure 9 is a surface plot that shows topographic details near the location of Smith's trench. This figure clearly shows the three-dimensional nature of the data we collected. The berm or ridge on the upslope side of the ditch is clearly discernable even though it has suffered considerably from erosion. The ditch depression is also evident even though sediment eroded from the berm has been deposited in the depression, shallowing it by nearly a meter from its prehistoric depth (Smith, 1961).

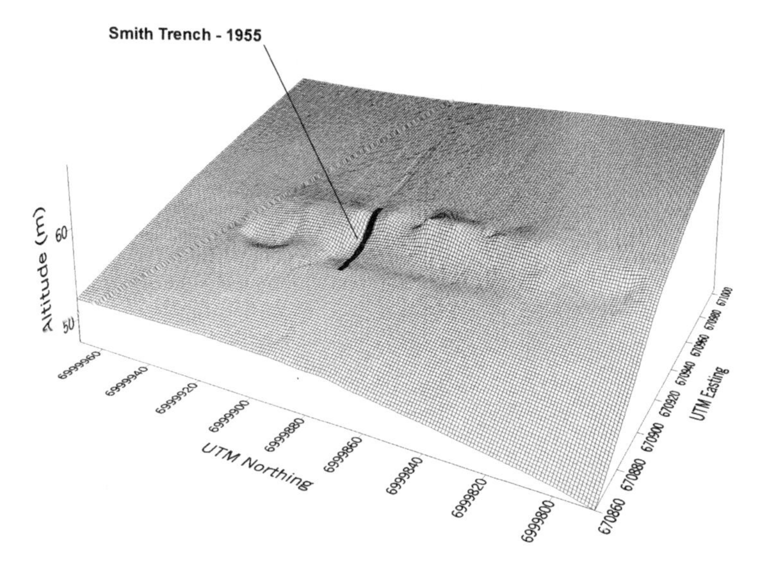

Figure 9. Surface plot of the southern end of the Poike Ditch showing the location of Smith's 1955-56 trench. The slight "bump" in the bottom of the ditch immediately north (or left) of Smith's trench is presumably the remains of the spoil pile from the trench excavation.

6. CONCLUSIONS

With our mapping effort complete, the topographic details of the Poike Ditch are now recorded digitally – preserving the morphology of the feature as it is today. Future erosion will undoubtedly continue to subdue the topographic expression of the ditch, but at least its condition at the end of the second millennium now has been documented. These efforts have also contributed incrementally to knowledge of the origins of the ditch. The map shows graphically that the ditch extends northward from the wave-cut bluffs of the southern coast of Poike, suggesting it is structurally controlled. We expect that our data will be of value to future researchers interested in the Poike Ditch and its origins, and to the Rapa Nui people as they strive to preserve the remarkable cultural heritage of their island.

ACKNOWLEDGEMENTS

This work could not have been accomplished without the assistance of the many participants in the Easter Island Interdisciplinary Expedition, most of whom either worked on the mapping project or visited us to offer their encouragement. We are grateful to Dr. John Loret for including the Poike Ditch mapping effort as one of the expedition's projects. Topcon America generously loaned us a GTS-311DG total station, and Trimble Navigation provided the Pro XL and GeoExplorer II GPS receivers used by the project. We are grateful to Dr. Daniel H. Mann and Dr. Christopher M. Stevenson for discussions of the island's paleoecology and prehistory. Marcos K. Rauch González and Rafael Rapu Haoa offered us their assistance and invaluable local knowledge. Finally, we are grateful to the people of Rapa Nui for their hospitality, and especially to the *vaqueros* of Poike who shared with us their knowledge of the Poike Ditch and watched our daily progress with silent amusement.

REFERENCES

Chubb, Lawrence J. 1933. Geology of the Galapagos, Cocos, and Easter Island. *Bernice P. Bishop Museum Bulletin* 110. Bernice P. Bishop Museum, Honolulu.

Englert, Sebastian. 1970. *Island at the Center of the World*. Charles Scribner's Sons, New York.

Thor Heyerdahl and Edwin N. Ferdon, Jr., editors. 1961. Reports of the Norwegian Archaeological Expedition to Easter Island and the East Pacific, Vol. 1, Archaeology of Easter Island. *Monographs of the School of American Research and the Museum of New Mexico* Number 24, Part 1. School of American Research, Santa Fe.

Knoche, Walter. 1925. *Die Osterinsel: Eine Zusammenfassung der chilenischen Osterinselexpedition des Jahres 1911.* Verlag des Wissen. Archivs von Chile, Concepcion.
Métraux, Alfred. 1971 [1940]. Ethnology of Easter Island. *Bernice P. Bishop Museum Bulletin* 160. Bernice P. Bishop Museum, Honolulu.
Routledge, Katherine. 1919. *The Mystery of Easter Island.* London.
Smith, Carlyle S. 1961. The Poike Ditch. *in* Reports of the Norwegian Archaeological Expedition to Easter Island and the East Pacific, Vol. 1, Archaeology of Easter Island. *Monographs of the School of American Research and the Museum of New Mexico* Number 24, Part 1. edited by Thor Heyerdahl and Edwin N. Ferdon, Jr. pp. 385-391. School of American Research, Santa Fe.
Smith, Carlyle S. 1990. The Poike Ditch in Retrospect. *Rapa Nui Journal* 4(3):33-37.
Thompson, W.J. 1889. Te Pito te Henua or Easter Island. *United States National Museum Annual Report.* pp. 447-552. U.S. Government Printing Office, Washington D.C.
Vives Solar. J.I. 1930. Orejas grandes y orejas chicas. *Revista Chilena de Historia y Geographica* 34:116-121.

Addendum

A summary of the work completed by scientists/specialists involved in the 1997-98 and '99 Interdisciplinary expeditions follows:

Dr. Daniel Mann, '97 and '98 – Paleobotanist Soil Geologist, University of Alaska: Obtained cores from the craters of Rano Raraku and Rano Roi (a record 55 feet, containing sediments dating back 80,999 years). Post analysis, working with Dr. Dorothy Peteet Lamont-Doherty Laboratories, Columbia University should determine: 1) resettlement climate fluctuations; 2) timing of forest clearance; 3) chronology of crop introductions; and 4) climate change over the last 1,000 years through the analysis of pollen, plant seeds and microfossils.

Drs. Robert Dunbar and Dave Mucciatore, '97 – Oceanography, Stanford University: Drilled coral cores at 7 offshore sites surrounding Easter Island. Core lengths should measure El Nino episodes over the past 500 years and give other historic data.

Drs. Warren Beck and George Burr, '97 – Geophysics, University of Arizona: 1) Coral core analysis for determining ocean's CO_2 absorption over the past 200 years; 2) Radio carbon dating of chronology since first habitation; 3) Cosmic ray C^{14} dating quarries.

Dr. Dennis Hubbard, '98 and '99 – Marine Geologist. Led a team of divers to do the first ecological underwater reconnaissance of Easter Island's sub-tidal marine environment to depths of 130 feet. The purpose was to survey the underwater terrain to determine its capability to support coral and

Easter Island, Edited by John Loret and John T. Tanacredi
Kluwer Academic/Plenum Publishers, New York, 2003

fish. They discovered the first live coral reef on Easter Island off the southwest promontory of Rano Kau.

Dr. Blaine Cliver, '98 and '99, U.S. National Park Service. Used photogrammatry to determine rates of erosion of the many petrographs, Moai and Ahu sites due to weather, animal (horse and cattle) activity and human impact. Dr. Cliver is a cultural historian responsible for protection of World Heritage sites for the U.S. Department of the Interior.

Dr. Richard Rainer, '98 – Geoarcheologist. Mapped the entire Poike ditch. Using GPS, Dr. Rainer was able to plot more than 2200 points. With this detailed survey we can begin to systematically excavate the Poike ditch, which is important to the agricultural system of the ancient Rapa Nui culture.

Dr. Andrew Merriwether '98 – Medical Anthropologist, University of Michigan: A human genetic specialist, Dr. Merriwether will be analyzing old human bones collected from ancient burial sites for the mitrochondrial DNA signature. Hopefully this information will enable us to better understand the origins of the ancient Easter Islanders.

Mr. Kevin Buckley and Terry Savage '98 – past Superintendents of the Gateway National Recreation Area, U.S. National Park Service, working with Chilean and Easter Island Park Service (CONAF), have developed a plan for Easter Island to better manage the Island's park land, and better preserve its monuments and cultural heritage. Approximately 80% of Easter Island is parkland and is designated a World Heritage Site. It is estimated that tourists will exceed 50,000 per year by the year 2005 and will continue to increase.

Dr. John T. Tanacredi '98 and Expedition leader in 1999 is a Research Ecologist retired from 25 years with the U.S. National Park Service. He is presently the Chairman of the Department of Earth and Marine Science at Dowling College, LI, Oakdale, New York. He coordinated the first comprehensive survey of wave pool invertebrate communities along the shoreline of Easter Island in cooperation with The American Museum of Natural History, Department of Invertebrate Zoology where he holds a Research Associate position.

Dr. Chris Boyko '99 – Department of Invertebrate Zoology, American Museum of Natural History, conducted invertebrate taxonomic identifications of the some 4000 plus specimens brought back on the

1998 – '99 expeditions to Easter Island. This is the first comprehensive invertebrate reference collection at AMNH for Easter Island.

Dr. Dorothy Peteet is a paleobotanist at Lamont-Doherty Earth Institute labs of Columbia University. (see previous notes under D. Mann).

Ms. Lindley Kirksey has conducted post-graduate work in existential psychoanalysis. She is currently a literary agent and reviews books for the Explorers Journal. Lindley is a Fellow of The Explorers Club and is a Fellow of the Royal Geographical Society.

Easter Island Expedition Participants 1996-1999

The Science Museum of Long Island, The Explorer's Club and the National Park Service, Gateway National Recreation Area

Jennifer Arnold '97
Ronald Asadorian '97
Norman Baker '97 '98
Warren Beck, Ph.D '97
Christopher Boyko, Ph.D '99
Chad Brayman '97
Kevin Buckley '98
Gerald Bunting '98
George Burr, Ph.d '97
Julianne Chase, Ph.D. '96
Blaine Cliver '98 '99
Toby Curtis '98
Sylvia Earle, Ph.D. '96
James Edwards, M.D. '97 '98
Joan Friedman '98
Michelle Garcia '96 '97 '98 '99
Patricia Galappo '98
Robert Grunder '97
Thomas Hall '97
Robert Hemm '96 '97 '98
Lori Hewitt '98
Dennis Hubbard, Ph.D '98 '99
Martin Isler '97
Natalie Isler '97
Lt. William C. Kempner '97
Virginia Killorin '98
Lindley Kirksey '97
Emily Loose '98
Elissa Loret '97
John Loret, Ph.D '96 '97 '98
Daniel McCann '98
Daniel Mann, PhD '97 '98
Ellen Marsh '99
Susan Marshall '98
Michelle K. Mass, M.D. '97 '98
Marcello Mendez '97 '98
Jane Montgomery '97
David Mucciarone, Ph.D. '97
Jessie Nestor '97
Alice Thor-Pianfitti, Ph.D. '98
Richard Pekilney '98
Jackie Quillen'98
Sergio Rapu '97 '98 '99
Rick Reanier, Ph.D. '98 '99
Susan Reanier '99
Priscilla Ridgway '97
Terry Savage '98
James Smyth '97
Rebecca Tait '97
John T. Tanacredi, Ph.D '98 '99
Elisabeth Taylor '96
Henry Tonnemacher '98 '99
Peter Vabula '97
Gary Warren '98
Brenda Tait-White '97
Joan Wiehl '97
Holly Williams '97

Index